# TILLERS

# TILLERS

## An Oral History of Family Farms in California

## Ann Foley Scheuring

**PRAEGER**

PRAEGER SPECIAL STUDIES • PRAEGER SCIENTIFIC

**Library of Congress Cataloging in Publication Data**

Scheuring, Ann Foley.
Tillers: an oral history of family farms in California.

1. Family farms—California—History. I. Title.
HD176.U6C27    1983        338.1'6        83-4179

ISBN: 0-03063796-1

Published in 1983 by Praeger Publishers
CBS Educational and Professional Publishing
a Division of CBS Inc.
521 Fifth Avenue, New York, NY 10175 USA
© by Ann Foley Scheuring

*All rights reserved*

3456789 052 987654321

Printed in the United States of America
on acid-free paper

**tiller**
- n. one that tills: husbandman, cultivator, plowman
- n. sprout, stalk; esp: one from the base of a plant or from the axils of its lower leaves

Webster's Third New
International Dictionary

# Preface and Acknowledgments

The reasons I undertook this project were in part personal and in part professional. I was once a city girl who married a farmer, and my education about agriculture took place almost by osmosis, in northern Illinois during the late 1960s. Within a very few years I became aware that many things on the farm were changing so rapidly that what I learned became outdated very soon. Among other things, we grew hand-picked tomatoes for soup manufacture in Chicago. By 1972 the tomato-processing industry had shifted dramatically westward because of harvesting mechanization in California, and we went out of that business. Listening to my in-laws reminisce, I knew that the changes they had experienced over time were even more striking because their perspectives backward were much longer.

As most people know, farming is now less a way of life than it is a complex and challenging business. Through many kinds of technological development on the farm, our views of agriculture have been altered, and a much broader social evolution has ultimately been a result. Many of us who have seen such change want to make connections between the past and the future, perhaps to ascertain the meaning of the moment in which we live.

This need for connections is not, of course, exclusive to agriculture. Certainly Henry Ford and his peers could scarcely have foreseen how the nation would be transformed after the invention of the horseless carriage. As a writer I am struck by the possible effects on language that may be brought about by computerized word processing systems. No one yet knows how the old art of putting thoughts into words will be influenced by the new capabilities — and limitations — inherent in computer information storage. In nearly any area, technological change leads to other kinds of change.

By the year 1975 my family and I had moved to California, where I became employed at the agriculturally oriented University of California campus in Davis. In 1978 I was a research associate on a project studying the effect on farm workers of the adoption of the electronic sorter in the tomato industry. Along the way I saw mountains of statistical research on agriculture; indeed, the collection of statistics on agriculture by federal, state, and local agencies is staggering. Students of change are thus able to interpret history with reference to many quantitative measures. The literature on agricultural development tends to stress numbers — indexes of productivity and consumer prices, reports of "inputs" and "outputs," collocations of figures on farm income and employment. The seduction of statistics leads to the econometric models that are the stuff of existence for agricultural economists. Although such quantitative surveys and analyses can be very informative, they tend to be narrowly focused and to emphasize mainly questions of material gain or loss. But numbers have no flesh, and nowhere in the statistical records is there a clue to how people react subjectively to change.

About this same time I read Ronald Blythe's <u>Akenfield</u>, a superb collection of interviews in a village in Suffolk, England, and I was profoundly impressed with its sociological significance as well as its literary merit. Since I had wanted to document the experience of change on family farms in California, I proposed to undertake a project in oral history to record farmers' views of themselves over time.

Thus began my interviews with selected families in the winter of 1978-79, through the University of California Agricultural Experiment Station, under the sponsorship of Dr. Orville E. Thompson of the Department of Applied Behavioral Sciences. Months of interviewing culminated in nearly a thousand pages of transcripts, which subsequently were distilled into a pair of journal articles describing the themes of the interviews.

Although by 1980 I had moved on to another project, I was unwilling to forget the treasures buried in the thousand pages of transcript. Instinctively I felt that the material was of interest and value in itself. Sporadically I kept working the transcripts toward some shape that might engage a reader. The laborious process of culling, selecting, and editing went on through many drafts while I queried publishers for their interest. Finally, in 1982 the much-reduced manuscript found a home.

Since the book was developed out of material originally gathered under the auspices of the Agricultural Experiment Station project, many thanks are due to Orville Thompson. His support and encouragement were steady, even when the transcripts seemed unwieldy and incoherent, and his commitment to the value of oral history was unwavering.

To the families who shared of their time and reminiscences with a stranger, I express my appreciation; their stories will, I hope, be illuminating for others. Thanks are also due to Rose Charlton, patient transcriber of tapes; to Guy Whitlow, who typed several versions of the manuscript; and to many unnamed others who encouraged my efforts along the way. Valuable assistance in locating illustrative photographs was rendered by Don Kunitz and the staff of the Department of Special Collections, Shields Library, University of California at Davis, and by the staff of the San Joaquin County Historical Museum. Three of the photos appeared originally in <u>A Guidebook to California Agriculture</u> (Berkeley: University of California Press, 1983).

My most heartfelt thanks go to my husband David and his parents, who were the first to teach me most of what I know about farm families.

# Contents

|  | Page |
|---|---|
| PREFACE AND ACKNOWLEDGMENTS | vii |
| INTRODUCTION | xi |

## PART I

| | |
|---|---|
| ANOTHER TIME, ANOTHER WORLD: THE DIETZ FAMILY | 3 |

## PART II

| | |
|---|---|
| WIDE OPEN SPACES STILL: COLUSA COUNTY | 25 |
| BOB BATES, COUNTY AGENT | 28 |
| A POCKETFUL OF PENNIES: THE O'LEARY FAMILY | 35 |
| THE NEW BREED: THE BREIDENBACH FAMILY | 59 |
| ALTERNATIVES TO THE STATUS QUO: THE WEIDNER FAMILY | 77 |
| SOME STAYED HOME: THE SAVELY FAMILY | 109 |

## PART III

| | |
|---|---|
| THE SPREADING URBAN WORLD: STANISLAUS COUNTY | 137 |
| PORTER WOODS, COUNTY AGENT | 140 |
| A VIEW OF THE ORCHARD: THE LIND FAMILY | 147 |
| THE MINISTER'S EGGS: THE LOWE FAMILY | 167 |
| PREVAILING WINDS: THE SCHOPPE FAMILY | 191 |
| STRUGGLE TO SURVIVE: THE WEBSTER FAMILY | 213 |
| THE RURAL ARISTOCRACY: THE MATHEWS FAMILY | 231 |
| THE VEGETABLE PEDDLER'S CHILDREN: THE YOKOI FAMILY | 257 |
| FIVE YEARS AFTER | 273 |
| EPILOGUE | 277 |
| ABOUT THE AUTHOR | 279 |

# Introduction

Few generations in history have seen such profound change in their worlds as those that have lived since 1900. This is particularly true for American farmers. Ask any 75-year-old (not so very old these days): Probably he or she grew up on a farm that ran by animal power, where feeding and harnessing the horses were a regular part of life, and the human back supplied all of the additional labor.

Consider American farming in the 1980s. In some places, at least, it is characterized by massive field machinery, complicated special equipment, evolving computer technology, and highly educated entrepreneurs. Many backbreaking chores have been eliminated by fossil fuel power and sophisticated engineering. Fertilizers and pesticides have become major supporting industries and provide some defense against disaster. Horses, if there are any, are pleasure animals.

Nowhere perhaps are the changes in farming wrought by technology more evident than in California. The most obvious and important of these developments is irrigation. Without water drawn from deep wells or brought from storage in mountain watersheds, much of California would still be dryland-farmed and limited to crops grown from winter rainfall. Water — how and where to get it — is indeed the crucial issue for today's agriculture. Its distribution has been at the heart of the technological revolution that has shaped California farming.

This technological revolution has abetted the industrialization of California agriculture. Some huge farms here, particularly in the southern half of the state, produce enormous crops, and "agribusiness" is a term that was coined in California. Critics of the system have variously expressed regret and alarm over the implications of agricultural industrialization, ranging from concern for exploitation of farm labor to warnings of environmental damage from overcropping. In contrast, "family farmers" enjoy rather widespread sympathy (though some of it may be sentimental), because it is thought that they care more about their land and do less violence to the environment. Is there really such a difference between types of agriculture today?

Although in the popular imagination California agriculture is dominated by large corporate farms, the 1978 Census of Agriculture reported that over 80 percent of the state's farms are individual or family-run. In addition, family corporations, in which the majority of stock is owned by persons related by blood or marriage, constituted nearly 84 percent of the total number of farming corporations. While it is true that these figures do not reflect proportionately the actual control of farm acreage in California, they do reveal that "family farming" is still highly important in the agricultural economy. It is, nevertheless, widely conceded that the family farm in California, as elsewhere, is under great pressure.

A brief look at statistics tells us something about changes that have occurred on U.S. and California farms since World War II.

Farm Numbers and Size in the United States and California

| Year | United States | | California | |
|---|---|---|---|---|
| | Number of Farms | Average Size (acres) | Number of Farms | Average Size (acres) |
| 1940 | 6,102,417 | 175 | 132,658 | 230 |
| 1945 | 5,859,169 | 195 | 138,917 | 252 |
| 1950 | 5,388,437 | 216 | 137,917 | 267 |
| 1954 | 4,782,416 | 242 | 123,075 | 307 |
| 1959 | 3,710,503 | 303 | 99,274 | 372 |
| 1964 | 3,154,857 | 352 | 80,852 | 458 |
| 1969 | 2,730,250 | 389 | 77,875 | 454 |
| 1974 | 2,314,013 | 440 | 67,674 | 493 |
| 1978* | 2,479,866 | 416 | 81,863 | 407 |

*Changes in definition of "farm" and in methods of data collection account for this surprising increase in numbers in 1978.

Source: U.S. Bureau of the Census, Census of Agriculture, 1969, 1974, and 1978, preliminary.

The trend toward larger and fewer farms in both the nation and the state has prompted some reexamination of policies that may have contributed to the decline in numbers. The concentration of agricultural resources in fewer hands appears, in some eyes, to be undesirable. Some think if clashes with the Jeffersonian notion of democracy based on a population of farm freeholders: "the most valuable citizens . . . the most vigorous, the most independent . . . and tied to their country and wedded to its liberty and interests by the most lasting bonds . . ." (Thomas Jefferson's letter to John Jay, August 23, 1785). Others, more pragmatic, worry that economic concentration can ultimately lead to price-fixing for agricultural products.

Numbers of farms are, of course, not the whole story. While farm numbers have declined since World War II, agricultural production has greatly increased. The sheer volume of American farm yield today compared with 50 years ago is astonishing — indeed, the envy of the world. Per capita income of the farm population has increased substantially in the last 25 years, and physical assets (land, buildings, machinery, livestock, and stored crops) have more than tripled in value between 1960 and 1978. Even accounting for inflation, it is clear that attrition among farmers has resulted in better livings for many of those who remain.

But it is also clear that many farmers feel hard pressed to keep up with technology and markets. Farmers are notorious for "talking poor," but the rate of U.S. farm bankruptcies today is higher than at any time since the Great Depression.

This volume documents a series of farm family histories in California. They illustrate adaptation to technological and social change over a period of two or three generations. Because California agriculture today is highly diversified, the farm enterprises described vary in size and in commodities produced. Two counties are represented, one in northern and one in central California. Counties and county seats have been identified by their real names, but all other place names have been disguised.

The northern county is very rural, with little population growth in recent decades. It has numerous large ranches, little industry, and no towns of more than a few thousand inhabitants. Its political and social climate is conservative, its opportunities for young people (by their own judgment) relatively limited. The central county is a bustling place with a rapidly expanding metropolitan center and many subsidiary industries linked with agriculture. Many small towns are surrounded by many small farms. Much of the difference between the two counties can be accounted for by local soils and the distribution of natural waterways. Because it was more feasible, irrigation came early in one county, changing the nineteenth-century pattern of farming in a semiarid climate. Livestock and grain ranching has lingered much longer in the other county, although recently a large irrigation system has been completed that will bring, perhaps, a new kind of future.

The 11 families in this volume are farm owner-operators who have been in the same location for at least 40 years, some of them much longer. Several individuals speak in each family. Their collective reminiscences form a kind of family portrait over time, a multifaceted picture. All the characters who speak in <u>Tillers</u> have been given pseudonyms. Families agreed to participate in the original taped interviews with the understanding that this collection was meant to be of representative rather than strictly personal biographies.

Tillers as a title carries a double load of meaning: those who till the soil, and those sprouts that shoot up from the base of a plant, as in field crops like wheat, rice, and corn, thus enriching the yield of the field. The body of the book is divided into three sections, each preceded by an epigraph evoking a certain note. Part I is the story of a classic small farm family that feels bypassed in recent times by more aggressive operators and is uncertain about the future. Parts II and III are divided by locale. In Part II, set in the more rural northern county, several speakers comment on what they view alternately as stability or as stagnation. In Part III, set in the central county that has experienced rapid urbanization, farmers are concerned with land use issues, labor, and social problems related to change.

The families within each section have been juxtaposed for contrast and thematic development. Many of the original interviews have been greatly condensed. What remains is a record of experience and reflection unique to each individual but illustrative of some very broad trends in California farming. At the time of the interviews the oldest speaker was 86, the youngest 22. Thus we have a composite view of what it has been like to be a California farmer in this century: the evolution of crops with the development of irrigation and new varieties and techniques; the growth of some farms and the disappearance of others; the succession of immigrant groups in agriculture and the shifting mix of family and hired farm labor; the economic pressures that have forced farmers to adapt their practices; the increasing need for education and managerial training; the sometimes uneasy tension between the hold of tradition and the pull of innovation. There is a richness of illuminating detail here that goes beyond abstract statistics to explain the impact of change.

Some of these families have been very successful. Others have just survived. Each family has a unique set of relationships. Some are very quiet people who have not ranged far from home. Others are well-traveled, cosmopolitan, active in organizations and politics; they may live in the provinces but they are not provincial people. They serve as articulate spokesmen for their peers, and they know the world beyond the farm.

Some speakers are optimistic about the future, but many are not. They defend the efficiency of the family farm: "When you only have a small acreage you can give it all number-one, 100 percent care. Your ability to care for every acre that you farm decreases with the number of acres. The more you have other people do things, the less efficient your operation becomes." But they are anxious about its ability to withstand economic strains. "I worry about the future of the family farm. I have read reports predicting that by 1990 agriculture is going to be totally incorporatized. What kind of disasters are they going to create? The only reason anybody sells a farm is because he is economically forced to. Nobody leaves the farm because he feels there is a better life. We don't feel there *is* a better life. We are really frightened."

Nearly all of these families have changed their enterprises over the years as economic and other factors have influenced them: small dairies have folded up, poultrymen have become orchardists, peach growers have diversified, grain farmers have become row croppers. These families have been able to stay in farming because they have adapted with the times.

The times, however, have exerted increasing pressure for productivity and performance, and this has exacted its toll. Although almost all these families enjoy farming, many of them speak of extremely long hours, heavy financial obligations, worry and anxiety. In two families the younger generation has chosen not to pursue the family vocation of farming, even given relatively good financial opportunity to do so. Though the physical strain of farming has been lessened by mechanization, bulk handling, and herbicides, farmers are operating under greater monetary risk than in the past. Keeping up with new technologies is essential but the costs are high. Land price inflation has meanwhile enormously increased the value of farm estates, lifting them into higher tax brackets, and farmers worry about being able to pass their land on to their children.

Ideas of success in farming now also encompass more than simple viability. One farmer expresses lingering regrets that he has not been more ambitious. It is not enough to have a small, neat, carefully tended farm; he feels trapped in a treadmill to expand. Farming reflects the pressures felt on small business everywhere — "Get big or get out" — and success is often viewed quantitatively. Partly this is due to the erosion of income in an inflationary period, but it is also a sign of changing expectations. A quiet life on the farm, even for those who love it, is measured against images of success seen elsewhere.

These families are generally characterized by warm relationships, and several speak of the advantages of farming for marriage and child-rearing away from urban social pressures. Geographic closeness and the ability to transfer property have also kept the generations together.

Work patterns on the farm have changed, however, and so have some relationships within the family. Women contribute to the farm as they always have, but in different ways. Eighty-year-old Anna Dietz speaks of long hours cooking for harvesters; Genevieve Savely describes the endless tasks and isolation of farming even in the 1940s. Farm wives now perform fewer farm chores, but they are often employed off-farm.

The oldest farmers speak of doing men's work when they were 12 or 15; but children work less now than they used to, and not nearly so early. As farm tasks have become more complex and dangerous, children's labor is no longer needed, or, in fact, even usable. Has this diminished development of a sense of responsibility at an early age? Some think so and speak obliquely of a dying work ethic. But children are getting better educations, and they must.

Farming has become professionalized. Most young farmers go through college to learn management skills. Some earn graduate degrees. Modern farmers often attend professional meetings and take part in "in-service" education. They are no longer only local in their outlooks; many speakers describe participation in organizations at statewide or national levels. Farm decision making is often based on an informed and detailed economic rationale, and record-keeping and management ability are sometimes more important than physical strength or skill, or even botanical or biological knowledge.

As practioners, these farmers don't fit neatly into romantic agrarian stereotypes. They like what they are doing, but they are businessmen, and it's easier for them to talk economics than poetry. Nevertheless, they acknowledge the existence of non-economic values. For most of them, farming is still more than just a business. It is a calling. Says one son of his father, "If my dad were to work 40 hours a week, he'd get done on Wednesday." Another muses, "When a construction man quits for the weekend, his building is not going to get sick, or if it falls down, it's not going to be crippled. Diseases and things like that just don't take a holiday. With cattle you have to be with it everyday." The Weidner family, concerned that their rich soils have tightened up over time, are going back to less chemical methods of farming: "We feel that we're stewards of the land. We don't own it, but we have to leave it better than it was when we got it. All the wealth in the world has to originate from the soil. When we get into that concept of economics, we realize that we have to protect the soil. Keep it for ensuing generations so they will have wealth too. In the last five years we've been going more toward an organic kind of farming, to build the soil back up again." And there is pleasure in doing the job well. "I like to see things grow," says one tough, weatherbeaten 63-year-old. "It makes me proud if I raise a good crop."

A sense of place is still very strong for most farmers. One wife reflects on buying the family ranch from heirs who wanted to sell it after their mother's death: "We had to pay each one of them off. We bought all the cattle and the whole ranch. It was a real thing to do. One time we were driving . . . talking about what we should do. I said, 'Pat, if you buy that estate, it will kill you.' And Pat said, 'Mom, if I don't buy it, it will kill me.' So I knew how he felt. I thought, well, let Pat buy it and die happy. It was his whole life, I knew then."

These families talk about independence, self-sufficiency, and the cohesiveness of a family that works toward a common goal. They share a feeling of pride in productive work. Their lives are not trivial or wasteful, and there is nothing abstract or questionable about what they do for a living.

And so we have pictures of change and of continuity. Though these are California stories, they reveal attitudes and concerns not unique to this state, and they give insights into problems that farmers face everywhere. There is not much nostalgia here for old ways or picturesque customs, but there are reflections on cultural shifts. Inevitably, along with the recognition of many benefits, there are some regrets — the consciousness of intangible losses in communities and in individuals' ability to be independent. Economic forces that seem impersonal to some, conspiratorial to others, have forced farm families to adapt.

Perhaps the meaning of change must be defined at the personal level. But these families speak for many when they speak for themselves.

# TILLERS

# Part I

Landmark

The road wound back among the hills of mind
Rutted and worn, in a wagon with my father
Who wore a horsehide coat and knew the way
Toward home, I saw him and the tree together.

For me now fields are whirling in a wheel
And the spokes are many paths in all directions,
Each day I come to crossroads after dark,
No place to stay, no aunts, no close connections.

Calendars shed their leaves, mark down a time
When chrome danced brightly. The roadside tree is rotten,
I told a circling hawk, widen the gate
For the new machine, a landmark's soon forgotten.

You say the word, he mocked, I'm used to exile.
But the furrow's tongue never tells the harvest true,
When my engine saw had redesigned the landscape
For a tractor's path, the stump bled what I knew.

James Hearst

Reprinted from LIMITED VIEW by James Hearst, published by Swallow Press.

Pride shows in the face of a farmer as he inspects his sacks at grain harvest. About 1910. Reprinted with permission of the Department of Special Collections, University of California, Davis, Shields Library.

# ANOTHER TIME, ANOTHER WORLD
## The Dietz Family

Herman's huge gnarled old hands rest lightly on the sides of his afghan-covered armchair. The dim living room of his two-story farmhouse is almost Spartan, but very clean and tidy. He too is very clean, his work clothes spotless, his face astonishingly smooth for his 83 years, his blue eyes as clear and guileless as a child's. A simple, God-fearing, home-loving man, not much given to speech. Words come haltingly; his work-worn fingers move as if they would fill in the gaps if they could. Anna sits upright, almost primly, in her straight chair, feet planted firmly on the floor. A lively intelligence twinkles in her face. Although she defers to her husband, her wit is quicker, her reminiscences fuller.

HERMAN DIETZ, AGE 83

I was born in 1895, about ten miles from this farm here. My parents' families were both from Germany. My mother was only five years old when she came here. My father was around 20 when he came. I was born west of here in the foothills. It was pretty rugged then. The roads were bad, not even gravelled, but dirt roads. We had the horse and cart and buggy, and that was the only way to travel. In the rainy season, it was hard to get around. The roads would be almost impassable when you had a lot of rain. They have pavement now up there, not like here, but pretty good roads, and you can go through anytime. But up there, even today, they are isolated.
    The home that I was born in and lived in, my folks' house, and the neighbor's houses used to be someplace from a quarter

to a half mile apart. Most of those houses are all torn down and gone now. People do not live there anymore like they did. The older settlers have passed away, and the younger ones who work there go back and forth from town to do their farming. The ranches are the same, but people don't live on them so much. In Europe the farmers used to live in the village and go out to their farms. When they came to America, they wanted to live on their farms. Now it's going back the other way.

I had five sisters and one brother. When I was growing up, we had a small dairy and some grain. As a child, I had to work pretty hard. The children on a farm did everything. When we came home from school, we had chores to do — and plenty of them. We didn't have time to run around and get into mischief like they do nowadays. I milked cows, did the plowing, and helped with the seeding. At haying time I had to do that. I was only 15 when I started out working full time. Families were scattered around the countryside and children just had to go to work when they got big enough, so I forgot about high school. It was too far away to go, anyway. Just one month over seven years is all the schooling I ever had.

In November 1924 I married my first wife. She was a local girl; her folks lived on this farm and this is the home that her folks built. I met her working here, back and forth. After we was married for six months, the wife took sick with tuberculosis. She lived for four years, but she was bedridden most of the time, and then she passed away in February 1929.

My first wife had a sister and a brother, but I ended up farming her parents' place. The summer of 1924 was when I leased the farm from her father. He died just shortly after that, and I kept on farming the place. When the mother passed away a couple of years later the place was divided. There were 240 acres and each child got 80 acres. My first wife got this 80 acres with the house. I continued the lease with the others' places. I was all alone after my wife died and I wasn't interested in the house, but they persuaded me to stay in it, so I did. Then Anna and I were married in 1931, and Steve, our only child, was born in 1934, and we have been here ever since.

My father taught me to farm, one thing and another. I picked up things from him, but I really was out on my own early, and kind of grew up by myself. All I have acquired financially was pretty near what I did myself. Not much schooling, but learning from what I did myself, and seeing things. I think of Lawrence Welk — I have listened to his program for years. He had very little schooling too, but he stepped up in life. That is the way my life has been. I have just gone step by step, by myself. This country would be better off if the younger generation would be more that way. Nowadays, they are handed too much. It is better to struggle. You value everything more if you do it yourself.

Since I have been a farmer, maybe 60 years, there have been awful big changes. From the dry farming in those days to irrigation now. The first row crop that they raised around here by irrigation was sugar beets. Eventually they raised corn, milo, and tomatoes. But in the 1910s and 1920s it was all dryland farming, even alfalfa. The first irrigation that we put in was on my homeplace in 1922, the year my father passed away. My brother and I put it in. Now that was only for alfalfa — no row crops at that time at all.

I milked three cows here myself in the 1920s. I quit the milking, outside of the family cow, after Anna and I were married. It was too much work and it wasn't enough to make any real money. The worst time was in the Depression. Along about 1931, '32, '33 and '34, times were bad. That's when we were just married. An awful lot of California farms went under. We were able to survive because I worked hard. We made enough to pay our taxes and live modestly. But as far as anything else, we just didn't have it.

We always put in 10 and sometimes 12, 15 hours a day. Even now I run the harvester, not for myself but for my boy. I don't want any pay because I don't want to be bothered with it, but I like to work. In the summertime I am usually out 11, 12 hours every day. Even Steve, from a boy, would take a 20-minute nap in the afternoon, but I never do. When I am through eating lunch I am gone again. Of course, when evening comes, I don't do like a lot of people. They don't get at it in the morning, but they work late at night. I never went for that. When the evening comes, I am through. I usually get to bed by 9 o'clock, but I am ready to get up in the morning anytime after 4 o'clock. I never use an alarm clock. Steve asks me sometimes to turn the irrigation pumps off, and I don't mind doing it in the middle of the night. When you get as old as I am, you don't require as much sleep.

It takes an awful lot of money to farm now. It's easier for a small landholder to rent the land out. Farming is more difficult than it was 50 years ago. There is so much more equipment you got to have, and fertilizer that we didn't have at that time, and certified seeds. I am just not interested in all that now.

I always did ny own field work. I didn't very often hire help. Before Steve was born, my brother and I had equipment together. Sometimes in the winter, putting in the crops, we would have to get the ground ready after the rain and then plant it. We usually had one or two men to drive tractors then, because we had to get it done in a hurry before it really started raining. We worked 24 hours a day, we had men work at night, and we worked in the daytime. At that time it wasn't so hard to get local people to help. We only paid around 35 to 45 cents an hour. They worked 12-hour shifts.

Some of the old days were good days, and then again they weren't. When I was young, more small farms were scattered around, and people farmed their own land. Now many of the little farms are leased out to the big farmers who have taken over. I hate to see it. But the little fellow just can't compete. I wouldn't say the little farmers around here are folding up, exactly. They may still own their places, but they are coming to where they don't farm — they have gone and got a job and leased their small place out. They do lots better that way. I have a neighbor who retired from work down at the university farm. For a small farmer he is pretty well-to-do. He has about 200 acres of ground. I was just telling Steve the other day that this neighbor's farming, what he does now, spoils his vacations. They go to Mexico every year! I am about the only one still around here of the older people. Some of those small farms have been sold, and the houses have been torn down. There used to be a lot of neighbors, but not now.

We just made plans to transfer the farm to Steve and Nancy. We have 160 acres, 80 here and 80 across the road. We had our will made years ago, but we checked into it again with the attorney this past year. Inflation has raised the price of the land. The land is too high for what it actually produces. Buyers pay more than what it's worth in cash return. We found out if we left our will the way it was and just kept the place until we died, then on our deaths Steve would have to pay in the neighborhood of $50,000 just to keep this 160 acres from the government. The inheritance tax, you see. So we made a gift of the 80 acres across the road to Steve and Nancy. We paid the gift tax for that place for them. That was enough money, even so — $5,300 just to give it to them.

This 80 acres here, when we were married, was mine at that time. At that time, pretty near 50 years ago, the attorney says, "It doesn't make any difference. Your wife will get it." But today they told us we should put it in Anna's name too. Well, it cost me $2,200 just to do that. I don't mind paying county taxes, and I don't mind paying sales tax, and I don't mind paying an income tax. But when it comes to inheritance tax, I will object. The inheritance tax law is one of the things making it hard for small farmers. They can't afford it. It is getting to the place now where things are so high that a young fellow can't even get started. If he can't inherit something, he is going to be in trouble.

Steve looks at farming different than I did, and that's all right. I was always more cautious about trying this and that. I was always afraid. I had seen what happened to other farmers. Nowadays you just can't think about that too much. Steve is conservative, too — but he will look at things and go ahead more than I did. He is willing to take a few more chances. You really got to now.

I never worked just for money. I think, regardless of what you do, whether for yourself or for your neighbor or for a company, you are supposed to do your best and give them value received. And I think that they should pay you what your work is worth. Not this thing of trying to get by all the time.

A lot of people worry about the weather, but you know, that doesn't bother me. When you farm, you have to use your own judgment, and do the best you can. You got the weather to contend with, but I never lost any sleep over that. Even if it got so they could control the weather all the time, why goodness gracious, it would be terrible. People would want it different.

I remember when I was probably 15, 16 years old. The folks knew a man in town who ran a dry goods store. He came out to see them and me, and he asked me if I wanted to work in the store. But I didn't want to. Of course, they needed me at home, but I am awful glad I didn't go, because I have enjoyed being out in the open all my life, and everything about farming. Only one thing bothers me now as I get older, it's the cold. Not the heat, I can take that. I ran a harvester, the first one they bought around here, for 22 years, and I was always out in the open. It got as high as $117^\circ$ — there was a thermometer on the harvester. But it never really bothered me. But now, when this cold wind blows, that's the worst part on the farm, if you have to work out in it.

I never did care to travel. I always had to work so much in my life that I got so used to it, I can't quit. They tell me that's what keeps me going, but I don't know. But if you have been active all your life, you can't ever quit. I couldn't move to town. I don't know what I'd do. I have been happy being a farmer.

ANNA JOHNSON DIETZ, AGE 79

I was born in 1899 in Mayberry, California, a very small town. In those days farming was all around the town of Mayberry. In the high school there were only about 100 students, and probably 50 percent of them came from farms. They were always very much concerned about the weather. They would stand at the windows and look out to see if was raining — or if it was going to stop raining, going to go on raining, or (if they needed rain) if it would ever rain again. I made the remark at home that I certainly hoped that I would never come to the place where I had to depend on the weather for my living! My father was in a mercantile business, in partnership with another man at a dry goods store in town. During World War I, when there was an opening in the local bank, he became a teller first and later an assistant cashier in the bank.

I began teaching elementary school in 1919. I taught for two years in a very small country district in Alameda County, and then in the city of Alameda for ten years. So I did have the experience of earning my own living. My first teaching contract was $800 for a year, or $80 a month for ten months. I paid $40 for my room and board, and out of the other $40 I had to buy all my clothes and incidental expenses and save enough for the two months in the summertime when I wouldn't have any pay. The experience of 12 years of supporting myself and learning to make do with rather limited funds was very good preparation for becoming the wife of a farmer.

I met Herman while I was teaching in Oakland, and we courted for about a year. I wanted to complete my degree, and I had two units to go, so I went to summer school that summer. We were married on the first of September, 1931, just a few weeks after I got through.

When I first came up to Yolo County as a bride, my first sensation, I think, was of being a stranger in a strange land where everybody else seemed to know each other. Life was different from town, no question about that. But I liked it from the very first. I enjoyed the absence of the pressure of being on a schedule. I'm a fanatic about being on time — nothing disturbs me so much as to feel maybe I am going to be late, maybe I don't have enough time to do this or that. That freedom in farm life was very attractive to me.

I found that there were times when everything else was put aside in favor of the farm work. The very first summer after we were married, Herman and his brother and his brother-in-law together rented a harvester. It was the first time that they had done their own harvesting, they had always had a custom harvester come in. They not only did their own three places, but they did some harvesting on the outside to help pay the bills for the harvester. For six weeks we harvested grain continuously. There was only one man on the outside who was hired, and that was somebody to sew sacks. They all ate here, of course. My mother-in-law was not able to do the extra work, so for most of those six weeks I boarded the sack sewer. He stayed here all the time. It was three big meals every day. Usually it was 8:30 or after in the evening before they came in. We had no way to heat water except with the wood stove and the old water boiler. There was electricity in the house when I came, but the old electric stove wasn't very efficient. I used it only for top-of-the-stove cooking. Water was heated by pipes inside of the wood stove; there was a tank. We made a fire on the hottest summer days, early in the morning, to get that tank hot. I metered out water all day as carefully as I could, so there would be warm water to run upstairs at night for Herman to have a bath at the end of the dusty, dirty day of harvesting. I would heat my dishwater on the electric stove in a big tea kettle.

I never did a lot of outside work on the farm. I thought when I first came, before I had family, that I would try doing some things, maybe learn to milk cows. But Herman's answer to me was, "Don't learn how and then you will never have to do it." Along about now, I am just as glad I didn't know how. My role has been pretty much secondary, supportive, all the way.

We had a garden over the years, but I haven't done the work in it. I have arthritis; I had back trouble as far back as high school. No one paid very much attention to it then -- your back just hurt. I got ambitious, the second or third year we were married, before Steve was born, and tried to do some work in the garden, and overdid it, and then I had to have some treatments on my back. After that, Herman said, "Let the outside work alone. You have work to do in the house." When he had time, he did what was done. He did a good deal more then than he does now -- he always said, "When I am too old to do farmwork, then I will have a vegetable garden and raise flowers." But that will never happen -- he will never stop farm work! We had corn and tomatoes in the garden before we had fields of tomatoes around us, and some other vegetables. Steve took care of the garden while he was in high school. I took care of the chickens, gathering the eggs and feeding the hens.

We had sheep a good many years after I was up here. Herman would be very busy during the lambing season. I often had lambs in the kitchen in boxes or up on the old wood stove, wiping them off and trying to get them warmed up. We had lamb bottles stored in the basement that came out every fall. I worked with the little lambs, trying to get them to take some nourishment, because there would be cold weather or sometimes a mother that had twins would only want to take one of the lambs. We tried to keep them alive until they could be given to mothers who had lost lambs, or perhaps their own mother would take them back in a day or two. I ran a lamb nursery inside here, although I never was out with the sheep.

Herman had about 100 ewes. It was part of the farming operation. Of course, farming itself was so much simpler in those days. You had a busy season in the fall and winter getting the ground ready and getting the grain sowed, and then a very busy harvest season. But in between, time was not quite so rushing. The sheep kind of filled in then, and also helped fill in some of the income.

But level ground, such as we have, is not the ideal place for raising sheep. With our wet winters, the ground was not very well drained, and the corrals would become muddy. It was hard to keep the sheep in good condition. Herman had the barn fixed up to shelter and feed them. That was quite a big task, morning and night. He bought alfalfa when we didn't have any. He would put it through a chopper in the barn, and then into the mangers. He had small mangers, too, special for the lambs, where they could get through and the mothers couldn't, so they could get a chance to eat as soon as they were old enough. It always worried him when the barn would begin to get muddy — the sheep would be outside, and they would come in all wet when it was feeding time, shaking off the moisture from their wool. That distressed him a great deal because he was very anxious to take good care of his stock. He had a great respect for livestock and tried his best, but it was rather difficult.

All during the time he raised sheep, Herman refused to eat lamb meat, so we didn't kill any lambs. He's learned since then to eat lamb, but at that time he considered them, well, kind of dirty. Nobody in his family had ever raised any sheep. They had some hogs when he grew up, and they smoked pork and cured it, and it would last without refrigeration, so they would have pork the year round. But lamb and mutton weren't that way. His family just didn't grow up on it.

In the 1930s during the Depression there was one year when we sold barley for 41 cents for 100 pounds. It must have cost at least double that to raise the crop. We borrowed money that year. There had been a very dry year and there was nearly no crop the year we were married. And then the prices went down besides. It was in 1932 that we got the lowest price. In 1934 Steve was born. I remember this distinctly because we were going to have a $100 bill for having that baby, $50 for the doctor and $50 for the hospital. It was a very big item! He was born at the end of August. Before that, when birth was impending, the grain buyer had told us we had extra good barley in one field. He said, "Don't sell that for under $1 for 100 pounds. We can get you that." That was very encouraging. Actually we got $1.02, and things began to look up from then on. And that paid for Stevie too!

I don't keep track of the years exactly, but by the 1940s was the war, and there were some increases in farm prices by then. Things were a little better. The first thing we did was pay off what money we had borrowed during the 1930s and get clear.

Our principle has always been that we don't buy anything that we can't pay for. You don't have a monthly income on the farm. You either have the cash to pay for something or you don't. You don't buy things on credit; you do without. When we wanted things, we planned. Herman was planning when we were married to do some remodeling on the house, but it was ten years or more before he was able to do anything at all. And then he did all the work himself. The only thing he ever hired to be done inside the house was plastering. We simply just waited for things, that's all.

Herman is a very conservative farmer and a very conservative man, in every way. He is very far removed from being a gambler — except insofar as farming (or any real business) is a gamble all the way through! He prefers, very definitely, to have a small income or a small profit, and be reasonably sure of it, rather than to take chances. He is just not a chance-taker. It bothers him terribly to lose any money if he feels that it was poor judgment or poor operation of any kind. That goes back to his early life, and mine was akin to it.

We had an old saying when I was growing up:

> "Use it up,
> Wear it out,
> Make it do—
> Do without."

And we lived by that as much as we could. Money came hard. You worked for it. Anything that you bought, you felt had to be worth what you were paying. Then you had to take care of it. You just didn't throw things away, or discard them because you were tired of them; you wore them out. If you outgrew clothes, then somebody else (friends or neighbors or some destitute family) would get them. My mother made everything. I never had a coat or anything except shoes that was bought in the store until I was high school age. My mother and father believed in quality, though not necessarily very much of it. Herman is the same way: If you are going to buy something, wait until you can buy good quality and then make it last. I still am a waste-not, want-not person, and so is he.

After Stevie got old enough, the summer he was 10, he had chores to do that he was definitely responsible for. He was not supposed to be told that they had to be done. He was supposed to take over the responsibility, and he was promised that he would be paid. That first year after he was ten, when the harvest season was over and the grain was sold, his father asked him which he wanted, money or some new ewe lambs. He took the lambs. He was given ten of them for his own. Every year after that, Steve kept some of his ewe lambs for replenishing his flock, and he would dispose of the older ones. The boy was then in business with his father until he got through high school and went to college.

The sheep were right here on the homeplace. We were still dry-farming. Instead of alternating crops every two or three years, a field would be pastured and the sheep would use that. Of course they also had the grain stubble. Everything was fenced then. On the three acres north of the house, Herman never wanted to have a flammable crop that could burn, because it was so close to the buildings. That's where the mothers with their small lambs were, until the weather got bad. When it came time to shear, Steve's lambs were kept separate from the others. He had to pay for shearing of his sheep himself. He elected most of the time to keep the ewe lambs, to build up his part of the flock. Then he would have a certain percentage of the flock, first 10 percent, then 20 percent, and so on. When he got up to 50 percent, his father said, "That's as far as you go, to have half of these sheep." From then on, it was strictly a 50/50 proposition. He belonged to the Future Farmers in high school and had that as his project. They worked out a contract that he would do certain parts of the work, and that would pay for the board of his sheep. By that time they had some alfalfa every year, and irrigation, so all summer long Steve would mow and rake the alfalfa and help with the irrigating.

When he went into the service his father kept the sheep for a while, but then it got to be more and more of a task. When Steve came back from the army the sheep were sold, because in the meantime he had developed some allergies and he got hay fever and skin problems from the hay dust. He could handle the green alfalfa in the field, but not the dried hay for feeding. I think there were close to 200 sheep when they finally got rid of the whole flock. Then they started to take out the fences. Everybody else was doing that too, because it made the farming easier, by keeping down the weeds.

My chief contribution to the farm has always been taking care of my husband — providing him with three meals a day, keeping his clothes washed, ironed, and mended, and his house in order, plus giving him companionship. Really all that has ever been asked of me is to run the house, and have the meals ready on time. Wives of farmers nowadays are different than in my day, because life is different. Most of them have more education, and some have been trained for business careers or professions, and so have developed interests of their own. We have more equipment to do our work and transportation to get around and do things. Life is easier than 50 years ago. Wives today are not confined to the farm as much as when I first came up here, and to an even greater extent before that. That's good. There is more to life than just earning a living. Now there is more room for participation and enjoyment and contacts on the outside, to make a life complete.

There was a time when I said to Steve, "For goodness sake, bring up your boys to be something else besides a farmer, because it is too tough now to get into it with all the expenses." Both of us even felt for a while that perhaps we had been mistaken in encouraging Steve as much as we did when he showed interest in farming. It hasn't been a very profitable venture from every standpoint. But if they are made that way, why, there is nothing else you can do.

Around the kitchen table in their modest rented home, Steve and Nancy talk of farming and family. Theirs is an affectionate relationship and a life filled with simple satisfactions. Even as Steve talks, however, articulate and calm, a shadow of worry crosses his tanned brow. Nancy, her face sympathetic, is aware of his concerns and conflicts. Here is a perfect example of the intelligent, industrious, caring small farmer — who feels somehow that he has missed the gold ring he should have reached for.

## STEVEN DIETZ, AGE 45

I was born in 1934 and have lived in Yolo county all my life. Our farm is about nine miles from town. All during the war years we went to town just twice a week: to church on Sunday and another time for grocery shopping. I grew up as an only child. My folks were ultraconservative. We didn't have the things that people have today. During the war there weren't many toys available, so I was forced to improvise. I made some of my own toys, mostly out of wood. I went to grade school in the tiny town of Lupine.

I was born, bred, and raised on a farm, and it is the only thing I really ever knew. I didn't associate with many other kids. Today, however, it is not uncommon for our eight-year-old to have a friend over at least once a week. The older two boys have played in Little League and Babe Ruth baseball, and athletics in the high school, and have taken part in many other activities that I never had the chance for. Their spectrum is much broader than mine was. My parents encouraged me to farm, though they left the thing up to me pretty much. But the thought that I would continue farming the family place was always there.

I graduated from Oakfield High School in 1952. I didn't graduate from college. I started a two-year program in plant science. Then friends and fraternity brothers talked me into switching to a degree program. But I had not had the preparation in high school for this, so I had to go back and pick up extra math. I switched late in the year and it was like carrying 100 pounds on your back up a 60-degree hill. It was a constant fight, and my grades were not good. I began to feel that to go into farming I could better use my time by getting the army out of the way. Possibly I could attend school after I got out of the service. But I didn't go back. I just came home and started farming the year after I got out.

In the military I was a welder in the engineering unit. The first two years I was stateside and the last year I spent in the South Pacific working on construction equipment. It was something I was really interested in. But like anybody else I counted the days and hours till I got out. My military experience didn't hurt me any. Later I built quite a bit of my farming equipment. We have a small farm, and it seems like it is getting smaller every day by comparison with others. There just isn't money to buy new machinery every time I need it. So some of the simpler tools I have built myself at quite a savings.

When I got out of the military in the summer of 1957 we started land preparation for crops the next year. I worked for neighboring farmers in the off season for the next year or so to help with the financial situation, but after I was on my own I did very little outside work.

Every once in a while the notion hits me to leave farming. I just came back from a meeting with my accountant this morning, reviewing the tax situation for this year. He said we made a lot of money (I don't know where it is), and we are going to have to pay a lot of taxes. I get discouraged. I told my wife on the way home that we could probably do better if we leased the land out and I got a job. Her counter was that I would never be satisfied doing something like that. And she is right, there is no question.

One thing I worry about is that all three of our boys, the older they get, the more interest they express in going into agriculture. But with inflation the way it is, outside people are buying up farmland at high prices, far beyond any return the farmer can get. Speculators are dealing with farmland as a commodity, much as you would gold or gems or something on the stock market. It's more difficult all the time to acquire land. I am very concerned that my sons will not be able to pursue something they are really interested in. I haven't been as progressive and outgoing as I might have been in the last 20 years. If I had been more risk-taking and more aggressive, our situation could have been different. I would be much better able to provide the boys with a start.

But people go bankrupt because they have taken too many risks. More all the time. It costs about $400 an acre to produce tomatoes now up to harvest. You have that amount of money tied up in growing the crop. Over $400 an acre — that is $40,000 for 100 acres. It is tough to take if you lose it. We lost our entire tomato crop in 1976 because of the cannery strike, and then later the rain. It's a touchy situation. I always felt that tomatoes were a high-risk crop, but it was okay to raise them as long as you had enough other crops (alfalfa, grain) that were more stable, to balance out the amount of money you could lose on the tomato acreage. It is no longer practical to think that way. Taxes and energy are going up. Tomatoes are the only thing that pay. We have to concentrate and intensify. We must have more acres of the better-paying crops to pay the bills. There is no other alternative.

We have a very small farm, just 160 acres. The boys work for me in the summertime and I hire very little other labor, except hoeing and thinning on our 35 acres of tomatoes. The rest we do ourselves. So the family is close-knit. But larger farmers have a regular labor force year around, and their kids are not involved. They go their own way. They have their activities and maybe even jobs in other places.

During the winter months, when the weather is bad, I work in the shop, and do a little bookkeeping. When the weather clears (February, March) we start field work on the tomatoes. Once they are planted and coming up, alfalfa hay starts the first part of April. Grain harvest starts in early June. This year I am thinking about double-cropping, which will intensify the work through the summer months, when it would normally slack off before tomato harvest in August and September. As soon as tomato harvest is complete, we start field work again. Land preparation for the following year usually continues at least through October. We plant grain in the last part of November and December. Then the cycle starts over again. The end of the year I am back in the shop.

I have never double-cropped before. I don't expect to make any money, really. If I can put the stalks and residue back into the ground, that will help the tomatoes the following year. With the way that tomatoes are harvested nowadays, with heavy trucks going back and forth across the fields, covering almost every square foot of ground, the ground gets more compacted each year. It is harder to work, harder to break up, to make a seedbed for the grain. We need to get some humus and organic matter back into the soil. The double crop is a soil-improvement technique, really. I'm just hoping I can break even on the actual growing costs.

Our farm is smaller than average for this area. I do more of the physical work myself and spend less time in management. I have less income from fewer crops. My farming has suffered somewhat from this, because my time is really more valuable than $4 an hour driving a tractor, yet I am forced to do most of that work myself.

The younger farmers who have come along since I got started are better educated than I was. Many are college graduates. I have felt a lack of knowledge about economics and management. I should have concentrated on these more than on mechanics and crop production when I was at college. Because that's the name of the game today — management, finance, economics. You can find out how to grow a crop, there is plenty of that advice.

But there aren't many people available who can provide sound financial advice.

Since I started farming 20 years ago, farms have grown larger, and equipment is bigger, with more horsepower, pulling wider hitches to cover more ground in less time. There is more dependence on hired labor, and more time spent, especially by the big farmers, overseeing things behind a desk. This is the way the world is going, not just in agriculture.

I have lagged behind in farming because I have been too conservative, worrying too much about saving money rather than spending a little more and having more volume and growth. I have a brother-in-law who is very progressive. He said to me once that regardless of how you feel, you have to more or less go along with the times in order to keep up. I wasn't sure at the time that he was right, but that has turned out 100 percent right. Well, maybe 80 or 90 percent. I wish I had realized a lot of things ten years ago that I realize now.

When my folks were farming, it was strictly a small grain operation. Way back in the early 1930s our farm was barley every year — burn the stubble, a few sheep to pasture the ground, and then barley again the next year. There was no use of commercial fertilizer then. During the war and after, we started to use commercial fertilizer. We started renting land out to neighbors for row crops in the early 1940s.

My folks had one of the first deep-well irrigating pumps in the area, and they were one of the first ones to level ground so it would irrigate. So in a sense they have been progressive, even ahead of their neighbors. But in other ways they fell behind. Later I used aluminum gated pipe for irrigating almost exclusively, because I do more of the irrigating myself and it is pure convenience — less work, fewer weeds to fight, less hassle with the ditches. We don't have too many acres so it isn't really prohibitive to buy the pipe. Pipe is water-saving. We have some spots that are sand-streaked and ashy-type soil where we would lose a third of the water in the ditch. But with aluminum pipes, every drop that comes out of the pump gets to the end of the pipe where it belongs.

I went into row crops myself. I had sugar beets for several years. Economics dictated that tomatoes would return the most, so we switched to tomatoes. I grew tomatoes starting in 1959 or 1960, during the hand-picked days. 1964, the last year for braceros [Mexican laborers who did seasonal farm work], was my last year for tomatoes. I was scared to death to depend on local domestics to hand-pick my tomatoes. Machines were not really dependable for harvesting at that time. For a combination of reasons, partly my own health, I dropped out of tomato growing. I didn't start again until 1975. A larger farmer here had been renting some of my ground to grow tomatoes on a year-by-year basis and agreed to harvest my crop for me. It has taken some of the headaches away.

We used to have problems with the braceros — whenever you have people, you have problems. I was sorry to see them go, but it was a change that had to come. The fewer people you have working for you, the fewer problems. Machines don't strike as long as you keep them in good running order and keep fueling the tank. When the braceros stopped coming, I felt there were no alternatives. I could not envision the winos being picked off the streets of the city and brought out each day to pick tomatoes. Of course, in those days, it didn't cost nearly as much to grow a crop as it does today. But even so, I felt it was too risky. My friends and neighbors around felt the same. Many of them dropped out of tomato production, at least for two or three years. Then, about 1967 or so, when the tomato machine was considerably more perfected, some of them went back into tomatoes. I should have gone back into it then too, with my own machinery. But today a new machine with electronic sorting costs $150,000. For my acreage it cannot be justified.

The canners prefer larger growers. The farmer who had been renting land from my father made it possible for me to get a contract with a new cannery. My small acreage went in, for contract purposes, as part of his. Had I wanted to start on my own at that time it would have been almost impossible. Most canners today want a minimum of about 240-250 acres per grower. There is less paperwork and bookkeeping and fewer people to deal with.

I had 35 acres of tomatoes last year. Next year I am doubling that. I hope to increase each year, and maybe get up to 150-200 acres. I would hope that I could increase some of my other crops too, for a little bit of stability. But my old theory of having enough other crops to break even on a tomato loss is just out the window. It isn't feasible anymore. If my sons come into farming with me, we will have to lease land. It would be the only way. And it would have to be pretty much tomatoes on the leased land.

The newer wheat and tomato varieties require more fertilizer than the varieties of years ago. Recommendations for nitrogen fertilization of wheat nowadays are at 100 or even more units per acre. Twenty years ago, 40 units was a great amount. In the 1960s I was applying probably a total of 80 units of nitrogen to the tomatoes. Now we have a total nitrogen input of maybe 120-150, depending on the type of soil, the particular variety, and the growing conditions. Of course, greater use of fertilizer results in higher production costs.

In a certain sense farming today is like a spiral. You have to keep bicycling faster and faster to stay in the same place.

You have to make more money because you have more costs. I used to figure $100-$150 an acre to grow tomatoes up to harvest. Now it is at least doubled, maybe tripled. Yet if you don't do all that, you fall farther and farther behind. The newer varieties are better producers, but they are also more finicky and fussy as far as growing procedures and operations are concerned. Used to be 25 tons to the acre was a good crop with the older tomato varieties. Now 25 tons is very little over breaking even. You have to get 35 to 40 tons to make money.

Twenty years ago I didn't borrow any money. But for the last ten years or so I have been a steady borrower for production loans, and a few times for an intermediate loan for equipment. I had a couple of loans from the Land Bank when we bought property. With some help from my parents and some money of my own, the actual borrowed amount has been low by comparison with other people, and it's always been paid back far ahead of time. But it takes more and more dollars to do the same thing that a few dollars used to do a few years ago.

My father always thought that one of the worst things you can do is to borrow money. I have an uncle who still can't believe that farmers borrow money for production now. His attitude is, if you have to borrow money to do something, you shouldn't do it. He is still thinking in the 1920s. To be sure, they came through the Depression and had things real tough at that time. They were fortunate to be able to hold onto their property. As things got better, they increased some in farm size. This particular uncle bought an 80-acre field for $10,000 in 1939. But he really had to be pushed to make the investment, he was scared to death. Well, he had it paid off completely in two years. The same piece of property today is probably worth $150,000.

I take pride in trying to farm well. I guess that anybody likes to top the next guy. With the newer varieties it is easier to get better yield. With the more modern machinery it is easier to do a better job. With herbicides you have cleaner fields and better looking crops. This, to me, is a great satisfaction. And I am my own boss. The work is hard and the hours are long, but I don't have to answer to anybody but myself. If I were just working in a factory and producing a certain part on an assembly line day after day, week after week, month after month, it would get tremendously boring. I like to do a variety of things, and I like to be outside. I wouldn't want to be cooped up in an office all of the time. I like physical exertion.

I am sure my children will approach farming differently than I have. They'd better, because they will be in big trouble if they don't! There are going to be fewer and fewer farms in the future, and they are going to become larger and larger. I think that eventually, down the road 20 or 30 years, there will be just a few large farms producing most of the food. The consumer is going to pay the bill, because the fewer people involved in the production of food, the easier it will be for them to get together and set prices — to hold commodities for what they want.

On the other hand, farmers are notoriously independent. Probably their greatest handicap is their inability to work together in cooperation. They produce food for a nation and for half the rest of the world. They could be in a whole lot better financial condition by collective bargaining. Unions bargain with employers for wages and salaries. But if 100 farmers get together and agree to hold a certain crop for a certain price, always a few will leave that meeting and think, "If these other guys hold, I can get in and get a better price . . . and I will sell right now and take the money." That is why it all falls apart. Maybe the fewer farmers the better able they will be to cooperate, the better educated they will be, the larger blocks of commodities and land they will control. Then it will be easier for them to work together.

A farmer always thinks next year will be better. Right now a lot of farmers are living poor and dying rich. Because of land price inflation, they have a great big dollar volume of assets, but the net return is almost nil compared with many other businesses. Yet when the farmer dies, if he hasn't done proper planning, he turns out to be a paper millionaire. Then the heirs have no money to pay the inheritance tax except by selling the farm. This is what I mean by living poor but dying rich. It's crazy.


NANCY DIETZ, AGE 45

I was born in Berkeley in 1934, but I finished school in the little town of Larkin. My mother was postmaster. After high school I attended the University of California at Davis. The College of Letters and Science was only a year old when I was a freshman. The Baggie Aggies were still in! At that time, girls didn't wear slacks. A "Baggie Aggie" was a girl who wore jeans to class if she took classes in livestock or was a pre-vet major (which was almost unheard of). I was a history major, so I wasn't a Baggie Aggie.

There were about 300 women to about 800 men in those days. In the early 1950s the Davis campus had only about 1,000 students, and it was a beautiful place to be. You knew everyone. Whether you knew their names or not, you spoke to them. That was the friendly "Cal Aggie" spirit.

I graduated and did a fifth year of student teaching, so I received my teaching credential. But I didn't teach. I just didn't have the confidence. During high school and college I had done summer office work for the grain elevator in Larkin, first weighing the grain trucks coming in and out during harvest and then bookkeeping. I went back to that after I graduated, and for a year or two following our marriage.

I met Steve first at a Sunday School picnic when we were both in eighth grade, but we didn't meet again until we were both students at Davis. We were married in 1957, after Steve finished his service in Guam. We have lived on the farm ever since. We have three boys — one is a college student, one a senior in high school, and one a third grader.

I am at Steve's beck and call on the farm. I try to be at hand if he says "Would you go to so-and-so and pick up a part?" It might be Oakfield or most anyplace. Or "Would you come out and drive a tractor while we unload pipe?" Or "I am going to spray today if the wind doesn't blow." I don't plan other things to do, I have to anticipate his needs. I don't drive a D-4 Caterpillar or anything like that, but I am there when he needs extra help. I do the bookkeeping from day to day, too. I do all the posting of the checks, expenses and income, and balance the books.

We have a good size garden. He plants it, and I hoe. In December Steve will get 400 or 500 onion sets, and then he wants to supply half the county with the produce in the summer! I do very little canning from the garden, but I have frozen almost every kind of food there is, and I make jams.

Steve was born when his dad was 39 and his mother 34, so there are really almost two generations between them. They were farming during the Depression and this sticks in their minds very vividly. But farming has changed so fast in even the last five years that it is quite difficult for them to grasp. They are maybe 20 years behind. Steve is more progressive. His parents' conservativism has been a holdback to him, and it has been frustrating at times.

Our sons are like Steve — deliberate. They especially have the feeling that what is right is right. They were raised in a mildly conservative atmosphere. I think this will remain with them, whether they pick up liberal ideas in college or not. They know what work is. Every time our college boy calls, he says, "What's dad doing now?" Of course he has been very involved in the business; he works all summer for long hours and has taken part in all the land preparation and the harvest. So his father's work is not something foreign to him at all. He has grown up with it and really understands. The big boys, since they were 11 or 12 years old, have driven tractors, and have done most things that a man would do. They were paid accordingly, but the pay was put into a savings account. Now it has come in handy for their college education.

I'd be pleased if our sons wanted to farm, except I don't see a great future in the small family farm. I wouldn't change what we are doing for the world, but we do get discouraged. I felt a little depression at times this fall. The tomato harvest wasn't the greatest. Last year we had a very good crop, but this year was ten tons less, about 26 tons to the acre. We didn't make what we had hoped. We have a boy in college and this is a great expense. Maybe I was depressed because the whole summer had ended and all the crops were in — just a feeling of anticlimax.

Steve gets wrought up sometimes, nervous or agitated about a crop. He keeps his cool pretty well, but under the surface there is stress — more all the time. He is very efficient with his time. He has to be, or he would be in debt. We haven't gotten bigger, but we have survived. And that is a triumph, that is the truth.

# Part II

The agricultural population produces the
bravest men, the most valiant soldiers,
and a class of citizens the least given
of all to evil designs.

> Pliny the Elder

Who has cleft a channel for the torrents
    of rain,
and a way for the thunderbolt,
to bring rain on a land where no man is,
    on the desert in which there is no man;
to satisfy the waste and desolate land
    and to make the ground put forth grass?

> Book of Job: 38, 25-27

Beyond the broad flat fields of Colusa County rise the mountains of the California Coast Ranges. Reprinted with permission from A Guidebook to California Agriculture, University of California Press.

# WIDE OPEN SPACES STILL
## Colusa County

A visitor to Colusa County in the 1860s wrote a description of the Colusa plains as "like a desert, and one of the most barren and inhospitable looking places to be imagined . . . principally occupied by bands of sheep . . . possessing no particular merit except ample room for improvement. . . . Little can be said of Colusa City, for there seems to be a general stagnation about the place. . . . There seem to be comfortable accommodations for a graveyard, and a long stay in the locality would render one a fit candidate for such a destination. . . ."

In the 1840s it is said that over 10,000 Indians lived in the area — or nearly as many people as today — but after the death of the great chief Sioc in 1852, the Indian population rapidly disappeared, decimated by white men's diseases and by ways inimical to their own culture. Grass was waist high in parts of the Sacramento Valley, and after the Gold Rush American settlers brought sheep in great drives from east of the Rockies, displacing some of the cattle that had been ranged by Mexican grantholders. Grain farming also began in the 1850s, increasing in importance throughout the "bonanza" years of the California wheat era. In the 1880s Colusa produced enormous winter wheat crops on huge ranches, and it was claimed in local newspapers that only two states in the nation produced more wheat than the county.

During the nineteenth century Colusa was home to a flourishing sheep population, which moved in bands of thousands up and down the open land throughout the valley. Drought and flood were periodic events, however, and some years saw disastrous losses in livestock. From the 1880s on, under pressure from grain growers, the range began to close. Fencing in the 1890s and early 1900s brought an end to some cattle and sheep operations.

Wheat production declined in California after world prices fell and the natural fertility of the land began to ebb under repeated cropping. The adobe soils in Colusa, where wheat had done well, proved after 1900 to be ideal for rice. Experiments with rice culture encouraged farmers to change to the crop that has now become the county's major commodity. Today Colusa leads California in rice production.

Two minor livestock industries attracted at least some farmers in the area for a while. From about 1910 up through the Depression, flocks of range-fed turkeys were raised for the San Francisco trade. Thriving on the grasshoppers that hatched on the sheep ranges in May and June, turkeys also fed on the residues of early rice harvests. By World War II the trend was to bigger operations, with new breeds being raised under confinement. With more specialized knowledge required by growers as margins between feed and turkey market prices grew slimmer, the turkey industry in California has since become concentrated on about half a dozen giant ranches farther south in the Central Valley.

Thousands of pigs from throughout northern California also used to be sold through middlemen to the garbage companies south of San Francisco, where they were fattened on garbage for sale to meat packers like Armour and Swift. Health and sanitation laws passed in the 1950s brought stricter meat inspection standards and prohibited feeding pigs on raw garbage, thus ending sales for such purposes. Health laws also ended the sale of weaner pigs to laboratories for experimental purposes, such as the development of antibiotics.

Almonds, now an important crop in Colusa County, appeared around the turn of the century. By 1917 Colusa had 5,000 acres in almonds. In 1912 an early subdivider in one area brought up several sections of former rangeland and planted 20-acre parcels to almonds, promoting the young orchards throughout Midwestern states and attracting investment buyers. Consolidation of these plots eventually took place, and in the 1950s a 50-acre orchard was probably average for the almond area. Thirty years later, almond production has expanded significantly, both in acreage, as irrigation has made it possible, and in yields, as research has improved varieties and techniques of cultivation. A 100-acre orchard is probably average today for a family living.

The Sacramento Valley is the major wintering area for migratory waterfowl on the Pacific Flyway, and Colusa County contains three of the four national wildlife refuges in the valley. Up to 3 million migrating ducks and geese rest here during winter months, and nearly 200 species of other birds have been recorded. Recreational hunting is popular among residents and visitors. In the late 1800s huge numbers of birds were slaughtered each year by commercial hunters who supplied markets in the cities. In 1889 four men in Colusa County killed 1,000 geese in one day. Such exploitation ended with game laws of the 1900s setting hunting seasons with strict limits on the birds an individual could bag. Illegal market hunters nevertheless continued to poach game birds up through the Depression. Federal agents during the 1940s finally brought an end to commercial killing when they captured and brought several hunters to sentencing in a federal penitentiary.

The county is settled into a prosperous agriculture now. There are still huge tracts of uninhabited land, some of it range for cattle or sheep, some of it riceland visited by migrating waterfowl attracted to its bounty. Driving along Interstate 5 at night, one can move for many miles without seeing any lights, especially to the west, where the coast ranges loom blackly up from the flat fields of the valley floor.

The ranches are not so enormous as they were during the wheat era, but they are still larger than the California average, and most of them are farmed by family dynasties. The towns are small, far apart, and not particularly affluent except for the county seat. The town of Colusa is a pleasant, leafy oasis along the Sacramento River, with attractive old homes harking back to days past. The Courthouse, built in 1861, is the second oldest courthouse still in active use in California.

Time has not quite stood still in the county, but the ambiance is of earlier times, when change was slow and life was ordered around the seasons. Some chafe at the conservative social climate, but others are glad for the stability and the freedom from the congestion, noise, and violence of the outer world. Colusa County is still a world unto itself.

# BOB BATES
## ─────o─────
# COUNTY AGENT

Bob has been director of Cooperative Extension in this county for 15 years. His office is modest, old-fashioned, in a small building on a side street in Colusa City. Bob talks from behind his desk in the compartment in back, which serves as his headquarters. His voice is deep and rich. He is a big blond man in his 50s, with a relaxed demeanor. He has plenty of sympathy for his farm constituency, but he also recognizes wider implications in what is happening in his county. His speech quickens as he begins to explore some questions.

Colusa County is bordered by the Coast Mountains on the west and the Sacramento River on the east. It is a small county for California — 800,000 acres — but the land slopes all the way from 8,000 feet in the coast range down to near sea level at the river. We've got three bands of farming, essentially, in the county. The older terrace soils lie right along the base of the hills, maybe 100 to 200 feet higher than the valley floor. That's where a lot of our row crops are — tomatoes and sugar beets — and many of our orchards. In the basin, the flattest and lowest part of the valley, are the poor soils. These old clay soils aren't worth a hoot; they hold water and they don't drain. Yet, as a consequence, we have 100,000 acres of some of the best rice ground there is. But there isn't much farming other than rice in the basin. Then, going east, as the river slows down in its flow through the valley, over a period of time it has flooded periodically over its banks. This has raised the soil level and the river area has been higher than the valley floor for many years. On those alluvial soils, lying right on both sides of the river, are some wonderful, very rich soils for orchards or for row crops. The soils are very deep and highly productive.

The county has 300,000 acres in agriculture. Livestock amounts to only 2 or 3 percent of the total agricultural production; rice is 40 to 50 percent; orchards, 15 to 20 percent; and row crops the rest of it.

The rice growers generally have very large acreages. Rice growers have to be large, because of the cost of the equipment. The harvesters cost over $100,000 now, and the bankout wagons are very expensive. But production is usually only about 50 to 55 sacks an acre, at $7 or $8 a sack. So rice requires large acreages but not very much labor.

Less ground is needed in other crops to make the same living. Orchards are more intensive and employ most of the farm labor around here. Row crops — tomatoes, sugar beets, and so on — are being grown more because of new ground going under irrigation. Tomatoes have become very profitable, though sugar beets are not good now. This county is the vine seed capital of the world. We grow more vine seeds than anybody else, for planting watermelon, cucumbers, and cantaloupe. It's a big industry here, though not very much heard of.

At one time the county had quite a few grape plantings, but prohibition and the phylloxera wiped out the grape industry. For a while people were getting interested in planting grapes again, but it never really materialized. Actually it's not a bad area for grapes, but the drawback is those tremendous amounts of MCPA spray, which is a 2-4-D material sprayed on all the rice fields by airplane. Grapes are very sensitive to 2-4-D, even at a distance. That is one factor, besides lack of wineries, that has kept the grape industry out.

In Colusa County the number of farms has declined some over the years, but not as much as in some other counties. We have around 600 farms. Conditions are fairly stable up here. The number of people has remained static. I don't think in the last ten years there have been more than 500 additional people. And yet the county is changing. The Mexicans are doing much of our work for us now — and a lot of them are becoming permanent citizens, too. If the population total is fairly stable, that means there must be an overall attrition of some of the Anglo farming people. I think that is happening. Other people associated with farming may be leaving as well. The number of small businesses related to farming is on the wane. As farms get larger or more diversified, the farmer buys bigger tractors. He goes where he can get the best deal. If he is going to buy three or four tractors at a time, chances are he will not buy them here in Colusa, he will buy from larger suppliers. A lot of our farm business is going out of the county. Farmers go down to the San Joaquin Valley, or even up to Oregon or Idaho, to buy their equipment. The people who have sold farm supplies, including fertilizer and everything else, have been Anglos. They are drifting away. As this happens, a sizable minority population is growing. In as small a population as Colusa County, shifts in minority percentages can be quite significant. The Spanish-speaking are probably better than 20 percent, way up from statistics of even ten years ago. This change is reflected in the school systems, in law enforcement, and in the burdens on taxpayers in the county. Some farm workers and their family members end up in our hospitals with very expensive operations. The problem is that agriculture enjoys the benefits of cheap, dependable farm labor without having to pay all the associated costs. The social costs are really subsidized by people other than farm employers.

Some Spanish-speaking farm workers move to the farm foreman level, but they certainly aren't going into farming, they aren't going into small business. I think most of them just want to come up here and do the best they can, and take their income back to Mexico. They are not too anxious to put down roots. There are some exceptions, of course. Some Mexican families do everything they can to involve their children in community activities. We see this in our 4-H program. But the large majority of Mexican parents seem to assume that their kids might as well go out and get a job as get an education.

There may be an increasing disparity between the rich and the poor in the county; it's how you look at the statistics. Colusa County has about the third highest per capita income in the state. But that's average per capita income. Taking the median income brings us down to 46th or 48th place among the 58 counties. This means we have a few very rich and a heck of a lot of poor people, relatively speaking. We do not have a large middle class. And the very rich tend to control the destiny of the poor. But this is no different from what has been the case all the way along, since the county came into existence. There were always gigantic ranches here, much bigger even than today.

We have had very low unemployment in the county. There is no industry where people can get jobs — so they just leave the area. As a result, our official unemployment figures are low, and the unemployed people are a problem some place else. That's one of the problems with the displacement of farm workers by machines. They go into the cities, and then they contribute to inner-city tensions because they're unemployed and sometimes very ill-adapted to city life.

But you can't convince a farmer that that's a reason he shouldn't buy a machine. His major concern is just trying to survive economically. All other industries have mechanized, so why can't he? Standard Oil has a refinery now that employs far fewer people, yet people don't attack Standard Oil the way they do agriculture. Agriculture is an easy victim for criticism — maybe because food is sort of an emotional issue. When you start talking about the food supply being threatened or somebody making a big profit off of what you are eating, people build up a lot of feeling.

Sometimes I think we have a management crisis looming for farm operations in this county. I see it in orchard crops, because there the Mexicans, and primarily the "wetbacks," do all our work for us. Growers tend to hire Mexican foremen because they know how to round up the necessary labor pretty fast, and it is very efficient. But because farms are using Mexican labor and especially Mexican foremen, a step has been eliminated for people going up the ladder into farm management positions. The Mexican foreman is very good for getting labor and for seeing that the work is carried out, but he is not generally recognized as having the capabilities for making management decisions.

As a consequence, we are already seeing a kind of crisis, particularly in orchards, where the father of the family dies and there are no kids to take over. The wife inherits the orchard, but what is she going to do? The family has to hire management to fill the vacuum. The number of farm management concerns is growing, because there are fewer people around who can make management decisions. It's nothing new. But now we are really beginning to feel it.

It has been a concern in rural counties that there is a sort of "brain drain" going on. People say we are losing our most talented folks from agricultural counties. For a doctor or a lawyer or even a teacher, the urban areas are where the best salaries are paid. If anybody is economically motivated, the rewards of living in a small community do not necessarily offset the opportunity to make more money in a larger one.

We are seeing some change in the population, however. Retired people are moving into the county because they like to hunt duck or pheasant, or do a little boat fishing on the river. They want to move out of the congested areas. We may see more retired people in this county in the future.

Colusa County has 50,000 to 60,000 acres of ground that will be irrigated soon, though we haven't put on water yet. The Tehama-Colusa Canal, a federal project, will bring water in all along the west side of the Sacramento Valley. So in all the counties along the way, there will be additional land going under irrigation in the relatively near future.

We will see an increase in orchards and row crops like tomatoes and sugar beets. It's good ground. Up to now they just haven't had the water to develop it, for that is an area where the wells have to be quite deep. The cost of pumping is high, and in some areas, even if they do have wells, the water quality is poor. Of course, farmers don't want to pump out of deep wells, anyway — they'd rather have surface-supplied water through subsidized irrigation systems.

The number of farms will increase, because much of the ground is now in large ownerships. Enforcement of the 160-acre limitation would mean a dividing of the big properties, either among the relatives in the family or through outright sales. I think it will be family members who will acquire the new farms. At least the canal might keep some sons in large farm families around here. In a lot of cases up here, there hasn't been room for all the kids to go into farming, because the farms haven't been profitable enough. But as you put land under irrigation, then the land becomes more profitable, and dividing the land then becomes feasible.

The great barriers to young people going into farming are lack of money and the availability of land. It's the old story everywhere: to get into farming you either have to inherit it or marry it. That is becoming increasingly true. I hardly know anyone who has started from scratch. But I can think of dozens who are in farming today because of family background. And yet the possibility of land ownership was one of the significant things about the frontier of our country as we pushed westward. People could relatively easily acquire land in the early days, but that situation has not existed for many, many years. The last people that I know of who got into farming without outside money or inheritance were the veterans who bought farms on the GI bills after World War II. Some of them, of course, were pretty successful.

Farmers are becoming very concerned over the image they have within the state. Because we are very dependent on rice in this county, we have a waste-disposal problem — we'll burn maybe 200,000 acres of straw after harvest. After a burn day, we can always read in "Letters to the Editor" in the Sacramento Bee just what a bunch of rats farmers are. We realize that burning causes a problem, and we are desperately trying to solve it, but there is no feasible way out of it right now.

The farmer has previously been able to do pretty well what he wanted to do, with limited governmental influence — but now the general public, through various organizations, is taking a hard look at farming, at pesticide practices and other things. Along the way, the government has played an increasing role in farming operations. The rice farmer, if it wasn't for government programs, wouldn't even be in the business. The farmer has to learn to adapt his thinking away from the individualism of the past.

Colusa is different from some valley towns. A group of old-timers has a great deal to say in how the city is run. They take pride in their community and are very active on the planning commission. They want the city to stay the way it is, but to maintain quality. They like the old houses, they want to preserve them. In fact, it is a status symbol to buy an old house and fix it up. Some houses go back 100 years. A unique thing about Colusa is that there is no public transportation in or out of town. No bus station, no railroad. It was built on the river and had barge traffic 70 or 80 years ago. River traffic has almost stopped now, but that history of isolation has contributed to the atmosphere of today.

Along the river, old families that go way back have stayed quite active in farming, some of them since before the turn of the century. They still do probably over half of the farming in the county. Maybe half a dozen families in this category are doing very well, and of course they are conservative about any changes in the county. Before social attitudes will change, new people will have to come in. Once I lived in a town that grew from about 10,000 to 25,000; then, all of a sudden, there was a whole new emergence of leadership. But for right now Colusa County is still being run by the old-time farm families.

This has been one of the few counties in the state, for example, where there is virtually no industry — not because people in general haven't wanted tomato canneries or other agriculture-related industries to come in here, but because the power structure hasn't wanted them. The farming population does not want unionization or a threat to the labor supply. They look at industrialization as a threat. This county doesn't even have a county housing authority. Poor people who need housing assistance have been kept out. Certain farmers even want to close down the one labor camp. They'll take care of the workers themselves privately, they say.

Now there are a lot of values implicit in all these things. Personally, I don't want any great change either — I'm very comfortable with the way things are. One of the great things about the county has always been our outdoor recreation: pheasant hunting, duck hunting, fishing. With new people coming, long-time residents find they don't have a place to hunt or fish anymore. There's no access to the river except in just a few places. The county is not what it used to be. If other changes were to go on, maybe some things would be taken away that would be irreplaceable. Maybe new problems would come in — there isn't much crime here now, for example. Compare Colusa with other counties and you can see that it doesn't suffer from the same problems. In some ways it may be lagging in social development (stores and entertainment and all that), but maybe that's good too. We might be ahead in other ways, you see.

The interesting thing is this: The people who control the destiny of the county aren't very many, but they're powerful and very influential. They try to keep the county the same as it's always been, yet at the same time they're the ones who can get out of the county. They have cabins up in the mountains, at Lake Tahoe and Lake Almanor, and over at Sea Ranch on the coast. They have airplanes. They can escape from the county when they want, while many of the people are really trapped. The poor are the ones who maybe want certain changes. But the people who now control the county's destiny, they want to keep it just exactly the way it is.

A 30-horse team, with a crew of 4 men, pulls the side-hill thresher of an earlier era. Reprinted with permission of the Department of Special Collections, University of California, Davis, Shields Library.

# A POCKETFUL OF PENNIES
## The O'Leary Family

It is a gray, wet winter day, and the little town of Plainview is sodden and wind-whipped, in sharp contrast to its dusty, sun-baked summers. On the edge of town are the O'Learys' three acres, an unpretentious older stucco house surrounded with a fence, animal sheds in back. Inside in the comfortable living room there is solid old furniture, decorative glassware, a grandfather's clock, and a new electric organ.

Pat O'Leary is small, wiry, graying, dressed in jeans, boots, and a pile-lined vest. He is slow of speech but enjoys being drawn out. His Irish background reveals itself in both his tenacious conservatism and his subdued twinkle. "How did you keep going in hard times?" He laughs. "Just stubborn, I guess." When he leaves the house, he climbs stiffly into a dented blue pickup of not too recent vintage and heads west into the hill country.

Helen is a foil to Pat: small, plump, outgoing, quick in speech with an occasional nervous, broken delivery. She moves around the room while talking, straightening some of her refinished Victorian furniture. She seems a generous soul, frank and full of life. Tears come to her eyes at moments when she talks about Pat.

Later, when Joe talks, the reticence of 68-year-old Pat becomes translated into the frustrated sullenness of a 22-year-old. He sprawls along a Victorian couch, ten-gallon cowboy hat tossed to the floor. He is not a talker, and has agreed to discuss his view of the ranch only because his mother asked him to. His face has not the quick intelligence of Helen's, but it is easy to see, stubborn and gruff as he is, what affection he has for her.

## PATRICK O'LEARY, AGE 68

My grandfather and grandmother both came from Dublin, Ireland. They were married in San Francisco in 1856 and went to Moore's Flat, a mining town in Nevada County. They had seven children, all girls except the last two, which were my father and his brother. My grandfather came over here to Elk Valley and bought 520 acres and paid for it in gold coin in 1880. That intrigues me, that gold coin. How he packed that I don't know, without getting robbed. He went back to Moore's Flat and was working in a placer mine when he contracted pneumonia and died in 1881, at the age of 60. He was 12 years older than my grandmother.

So my grandmother and her seven children moved to Elk Valley. They came in a horse-drawn vehicle called a spring wagon, with two horses and a few chickens and a cow. My dad rode a horse all the way. They made first camp the other side of Loganville somewhere. It took them three days. My grandmother had never been to Elk Valley before. There was a house and a barn up there when they came, and with her seven children my grandmother started ranching. Amazing.

The oldest of the girls was only around 16 or 17. My father was the oldest boy, and he was only eight. Anybody today in that kind of circumstance would be declared poverty-stricken, I suppose. My grandmother leased the range out. She milked cows and raised chickens and turkeys. She'd take butter and cheese to the mines at Sulfur Creek; they mined gold and silver up there, and they had a big hotel then, though there's nothing left hardly anymore now.

My father grew up in Elk Valley. He went to a little one-room school. The moment he got home from school he had to hook a harrow onto a horse and work some ground. He grew up working. The first job he had was pulling a header wagon for Herman Blythe, a farmer just north of the homeplace. All he had to do was drive. They brought the grain that way into the stationary thresher and pitched it off to the separators.

All the brothers and sisters lived on the ranch with my grandmother until she died — those that didn't get married and move away. My father's brother went to business college but he didn't stay there too long. They did farm together some after my grandmother died, but then my father bought the rest of them out, along about 1907.

There were quite a few small ranches in Elk Valley at that time, all cattle and dryland grain. I don't know how the ranchers acquired that land. I know a lady who has a map of old Colusa County about 1853, and on that map it shows that a fellow named Turner owned all the land in Elk Valley. He sold it off in smaller pieces to homesteaders, I suppose. My grandfather's land was originally acquired from a Mr. Grim — I have the original deed, and I found another deed to Mr. Grim before that. The legal description isn't written up the same, but it is the same property. I don't know if the original was a Spanish land grant or what. I was going to look it up, but it was something I never got around to doing.

My mother and dad got married in 1909. My mother came from a big Irish family too. Her brothers and sisters had some great names: Aloysius, Vincent de Paul, Francis and William Lawrence, Alexander, Ignatius, Gregory and Alphonso. Eight boys. Then there were five girls: my mother Rose, and Assumption, Patricia, Anita, and Virginia. My mother's folks had moved from out here somewhere on the plains to what they call Little Valley, just another ridge over from Elk Valley. They farmed in there and the children went to school down in the center of the valley. But I don't think my mother stayed up there very much. She stayed with her grandmother in Colusa and went to school in a convent. When she graduated, she taught school for a few years. My father and mother met because they had been neighbors, but by the time she got married, my mother's family was farming over in Sutter County. When my mother and father were married, he brought her back to Elk Valley.

There was more population in the valley then, before World War I, than there is now. At one time there were three little one-room schools up through ninth grade. The Elk Valley School closed before my time, though my father went there. Sulfur Creek School closed when I was little, and the Wilsonville School closed about 1930.

We have a buggy out here in the garage that they used to go back and forth in from the valley to town. Seventy-five years ago my father used to drive a four-horse team to pick up supplies. They could make it from the ranch in less than a day, but they'd have to stay overnight to load up supplies before going back. After harvest they'd give the mules or the horses a good rest and then they'd haul grain to town, come out one day and go back the next. They'd take 20 days to haul the grain into town. There was a railroad into Shepherd at one time and they'd haul over there too, 15 miles. They could make that 30-mile trip in one day with an eight-horse team. They'd drive cattle to the railroad too.

My mother and father had ten children. Two died in infancy. There were just two boys, me and one brother. I was born in 1911 and grew up on the ranch. And I went to the one-room school in Wilsonville. It is all very much different now than it was then. Practically all the buildings — the barn, the corrals — burned. There's no house on the homeplace now. If something burned, there wasn't anything you could do about it.

In the early days they used to say that was all meadow ground there in the valley. It's really marginal soil. With water it could produce, but there isn't much water. We have a reservoir up there now and we can irrigate some.

When I was a boy, maybe ten families were living in the valley. Now there's just one family living at Wilsonville, and one other man and is wife — he doesn't own anything, he just takes care. Nobody who's really active in ranching lives there. Some of the old families just died off, and the ranches got sold. It's too small a valley to support a big enough community to be a viable place to live. There used to be a post office at Wilsonville that was kind of a social place — people would go to pick up their mail and then go to the store and visit. But the post office closed along about 1923.

When I was growing up it was very seldom that anybody went on that road down Elk Valley. It was just two tracks and grass grew in the center. In the wintertime there was no traffic hardly at all, because it was all mud.

We did plowing and harvesting with mule teams or horses — 32 head on a harvester. We had some custom harvesting done. You'd pay so much for a span of mules or horses for the day. Then we had a good number of horses and mules on our own place, maybe 15 or 20, counting the young stock. We used animals for all that work for years. Finally, just after World War II, my father bought a tractor. But he still used teams on the harvester a couple years after he had the tractor. The neighbors he did custom harvesting for wanted him to use their teams, so he did.

Here is a picture of a 32-mule team hitched to a 16-foot grain combine. That's how they did it years ago. It took four men to harvest. One was the separator man, then the header tender, the driver, and the sack sewer. Mule teams were used until maybe 1920. They used draft horses, but not those huge teams, for 20 or more years after that. Of course, turning around that many animals was a trick. They couldn't get up to the corners quite as well. But a 32-mule team could harvest just as many acres in a day as we do with the machine harvester we have now.

For plowing, the teams would be broken up into sixes, eights, tens, sometimes twelves, with two abreast. There are six abreast in this picture, with the two leaders out in the front. See this man up here, sticking way out? He was the driver, up there on a seat that came way out over the front wheels. Sometimes they had what you call a jerk-line on the near leader, just one line. And sometimes they had what you call checks or lines on both of the leaders, way up to the front. You would try to get your smartest animals up in the front. To get them to obey a command, you used the jerk-line, and you'd holler, "Gee" or

"Haw," different syllables so they wouldn't get mixed up. Gee was to the right and Haw was to the left. If you wanted to go Gee, you jerked on the line that had the Gee string, fastened on a small strap to the collar, and it would throw their heads up, and they'd turn that way. And Haw, you just pulled on it easy. I never did that myself, but it must have been quite a skill.

Now in this picture there's a 12-mule team hauling Bartlett water out of Bartlett Spring. That's mineral water. The spring is about 20 miles back in the mountains. Bartlett was the fellow's name that discovered it. It was sold in the cities, all over. Mercury was mined down at Sulfur Springs for a while too. During the war the mine operated, but it is closed down now. Mercury has to get to a pretty good price before it's worth mining.

My parents raised turkeys. Just before Thanksgiving they would pick the turkeys and hang them out. Early next morning they would pack them in boxes and load them in the wagon, and my father would take them in to meet the train going to San Francisco to the commission houses.

My mother worked awfully hard. She would cook for the harvest crew (when she wasn't pregnant), and she did all the canning. She had a vegetable garden — turnips, carrots, radishes, tomatoes, and things like that. They planted some fruit trees but they didn't do very well. We would go to my grandparents' over by the river and get peaches, apricots, pears. Her brothers always came up to the ranch during hunting season, and they would always bring fruit.

When I was 12 or 13 I started deer hunting along the creek. I wasn't a very good shot! Our family used the game we killed, of course; about the only fresh meat we ever got was deer meat. We would kill hogs around Christmas, and then salt, cure, and smoke them and make sausages.

For years my father had hogs up there, probably about ten brood sows. We'd fatten those pigs out, and buyers would come along to the ranch after them. It was part social event — you would sit down and chew the rag with the buyers. We usually had to haul the pigs in to the railroad. We would wait until a moonlight night so they could drive at night, to be cool. It was usually in early fall. The pigs would be run on the stubble up to a certain point and then we'd get them in and feed them grain for a couple weeks. My dad had the pigs until they were almost a year old, and he'd be selling 30 or 40 at a time.

We had chickens too, for home, and then if they had an excess of eggs, why, my dad would bring them into town and sell them to the stores. If we had an excess of cream, my mother would make butter and bring that in and sell it. My dad also had cattle, of course, all this time, and put up a little alfalfa hay. All that was typical of Elk Valley ranching at the time.

The ranch was pretty much self-sufficient. We were able to get through weeks at a time without getting provisions. The women put up a lot of fruit in the summertime. They would buy field-run beans and store them, and potatoes, three or four 100-pound sacks. If we ran out, why, we'd take a trip and get more.

We had plenty of chores to do as kids: feed the chickens, feed the turkeys, milk the cows, water the garden. My grandmother's place had a dug well, but there wasn't much water in it. They drilled a well later down a ways from the ranch house. The wells down at the valley floor are better, they never go dry. I put a well down a few years ago right north of the house. It's 400 gallons a minute. You can sprinkle irrigate with that, though the salt content is a little bit high.

Ranching was fairly profitable in those days. Of course, we didn't have the things to buy for the home in those days as they have now, but we had hot and cold running water. Many of the ranchers didn't have that. Electricity didn't come there until about 1941.

Traveling back and forth to town was quite an undertaking — mud sometimes the whole way. The roads were all open ground, there were no fences. One time when my father came back from town I asked him how the road was. He said, "It's 100 yards wide and a foot deep." Sometimes the mud would clog up on the buggy and wagon wheels and they wouldn't turn. You had to get down there and shovel it off. Ranchers didn't market in the wintertime — unless it was turkeys or eggs, which could go in a lighter rig — they saved everything until the roads dried up.

The buyers for the cattle and hogs would come to the ranch, but the grain, we'd bring that in and put it in the warehouse. During the 1920s, when I was in my teens, my father was making a good enough living for our big family. We always had plenty to eat. We had a wood stove, a fireplace for heat. And with 10 or 12 families up there life was sociable. We had parties in the wintertime, card parties and dances.

But people started moving away from Elk Valley. My mother and father moved into town here in 1927 so we could go to high school. From then on, the population in the valley declined. People would come in and lease a place for a few years, and then leave, and someone else would come in. They'd all get discouraged. One reason it's so isolated is that there is a six-mile grade going in there. If you could go straight through, it would probably be a 15-minute drive or less. But you couldn't get a road in that way. Some of that Elk Valley road is real crooked. It scares some people. It is steep and drops off like that, in some places — a big precipice.

After I finished high school I took the two-year course at Davis. I graduated from a nondegree curriculum in 1934 and came back home.  That was the Depression, and more than once my father wondered how he was going to pay his taxes.  But he never ever borrowed money on the ranch, it was always liquid.  With turkeys and hogs, he had enough to keep the ranch going through the 1930s, and just before the war my father bought quite a bit of land.  Part of it was the Clinton place south of the homeplace.

My father was strict.  If you didn't ante up to what he said, you were in trouble.  In his younger days he worked dawn to dark.  But he always had a hired man.  When he moved to town, he kept someone at the ranch all the time.  In those days a hired man would be paid so much a day.  If it rained and he didn't work, he didn't get paid.  But he got his board and room.  Now you can't find people like that.  It has been a few years since I've had a full-time hired man.

I had turkeys up to about 20 years ago, though I think those commission houses were cheating everybody.  I raised hogs up until six years ago.  In rainy weather I would put the sows under cover to farrow.  If the pigs got rained on, they died.  In the summertime, you just let them go.  The only thing you had to do was provide shade for them.  They were all fenced in, though.  My dad said that the only way you could make money on hogs was to let them run on the neighbors!  But I found the neighbors objected!  I just fed the pigs barley.  Feeder pigs brought a good price — the city garbage people were buying pigs to fatten them on garbage.  I'd sell maybe a 100-pound feeder pig.  They used to like to buy them smaller than that.  But if you sold a smaller one, you'd have to get more than twice what you'd get per pound for a heavy one, to bring so much a head.  That's the way I'd figure it.

I phased out of turkeys and hogs because it got to be too much work, and labor went up so high you couldn't afford to keep anybody.  I don't think there's anybody who wants farm work nowadays.  I wouldn't have trouble finding a farm hand if I paid him $1,500 a month — but then he would want Saturdays and Sundays and double holidays off, and eight hours a day.  If a cow was having a calf out there and she needed help and it was after quitting time, he wouldn't look at her.  People are pickier than they used to be.

I have just cattle now.  We own about 6,000 acres all told, but 90 percent is rangeland not good for anything else.  The cattle business until a few years ago was number one for income in Colusa County.  Now it's third.  The rice business has taken over first.  All this alkali ground out here on the flats used to be just pasture for cattle.  After they found the correction for alkali, they put the land into rice.  There's more money in rice than in cattle, and it's more dependable.  Rice is just a few months' work in the spring, a few months in the fall.  The rest of the time is idle.

I kept my own cow herd up until the beginning of the drought. I was working with the Extension Service over 15 years on replacing heifers. We'd get a postweaning weight and put an identifying brand on them. At the beginning of the summer they were reweighed and we'd pick out the ones that gained the most. The gain plus the conformation score would make the heifer we'd save for a replacement. Extension Service thought the genetic factor would operate to upgrade the herd, but it didn't go as neat as they thought, even using bulls with a high rate of average daily gain (like 2.8 pounds would be a good gain for day of age). At the end of the year I'd sell the animals I didn't want to keep.

I had my herd up to over 300 at one time, but cattle prices went down and I cut back. When the drought came I sold down to 50, because feed was so high. There wasn't any grass on the range at all for two years. I am gradually building up again, buying and selling.

In the winter the cattle go out on the range. But toward the end of the spring, when it starts getting dryer, they start coming back in. Usually by summertime they are almost all in. We have to go out and get a few of them, but we don't have a big roundup. We still use horses for herding. The cattle stay in the valley in summer, on 135 acres of irrigated pasture. There is no water out there in the hills in the summertime, and they don't want to climb around when it is hot.

We brand them with a hot iron; I still do it myself. For a while we would rope and brand them. But now we separate the calves from the cows and from the corral we run them through a long, narrow chute into a squeeze. There is a lot of noise and dust. We brand and vaccinate them at the same time.

Seems like there are more diseases than there used to be. More parasites, flies. Cattle are transported farther now, and they carry diseases from other places. Before, when we bought cattle, it would be local. Now cattle come in from Idaho in a few hours, or even from Texas or Louisiana. We shipped a load of heifers not long ago and they went to a Nebraska feedlot in 30 hours. So they take some germs back there, and they bring some back here. Used to be the only thing we'd vaccinate for was Blackleg. Now we like to do IBR, BDB, anaplasmosis, rectos spirosis, and other things. Then there are parasites. We've had the horn fly, a little tiny fly, for years. But now they have a face fly that looks like a housefly, which swarms on the cattle's faces. If the cattle stayed in one place they wouldn't pick up all this. But we'll have it from now on.

We don't live on the ranch anymore, but I'm up there every day checking on things. It takes about 35 minutes to drive to the ranch from here. The old Clinton house is still there, but it's empty. I eat my lunch there.

Elk Valley is paradise for the hunting season. It's so out of the way. We have the wild hog, the quail, the dove, and the deer. People want to come and see us when it's deer season — we have dear friends then, you know! We did sell hunting rights for a while, but that can be a headache.

People do drive through there occasionally. The peak of the beauty is in the spring of the year when the wildflowers are in bloom. But we can't keep anything up there. We would have to stay right there and sleep with it, because we have been hit so hard. People have taken our furniture and our camper.

The busiest time of year is haying season. I was the first to have an automatic baler in the county. And of course we converted from horse-drawn haying equipment to power mowers. Being a horse lover I kind of hated to give them up, but the last time I used a span of horses, a neighbor came down and said he would rent me a tractor for $10 a day. I said, "That's cheaper than I can drive this team." I used horses up until about 1948, I guess, for haying and seeding grain. Through World War II with a four-horse team. That's later than most, but sometimes I'm a little bit behind!

The valley floor there is serpentine soil. You get down some places, you hit a hard substance, although not really a hardpan. Mainly we grow just wheat, and a little barley, on about 350 acres, and still all dryland because there's not enough water to irrigate. These new Mexican wheat varieties produce more, but I've had crop failures with the new varieties that I never had with the old ones. The new ones mature a little quicker, and frost gets them. Up there if you don't plant early, sometimes it starts raining and you'll never get on the ground. So if you plant early and you get an early rain, the wheat comes up and then heads out in April. And sometimes you get a frost in April, because the valley is about 1,300 feet.

We used to do straight Herefords, but then we got to crossbreeding Angus bulls on Hereford cows, and then their offspring with shorthorn bulls. No artificial insemination, just nature taking its course. We don't finish them out, just sell them as feeders. Years ago we used to grass fatten them but then the feedlots took over.

A month ago I went up to Round River and bought 115 steers that averaged 430 pounds, and the average price was 79 cents. But I was up to Cottonwood last Friday and some steers weighed 380, and they wanted 91-1/4. I didn't buy any. We had an auction-stockyard here in Plainview until about seven or eight years ago. But in 1971 the father died and left it to his three boys, and there just wasn't enough in it for three of them. It could go if they got all the cattle from Ukiah this way, but some winter months there'd be only maybe 100 head, and maybe 1,000 in the summertime. And then the Stockyards and Packers

Act said they had to have so much in reserve to cover checks, and lots of times they didn't have that. Now the Gurney Meat Plant closed down here, too, and they were the only ones in northern California that slaughtered lambs.

I sell my wheat right from the field. You estimate your tonnage and the Cargill representative checks if it meets government grade and specifications. He quotes you a price, and if it looks good, well, you just take it. There's nothing much else you can do unless you want to store it. First time I ever stored it, I could have got $5.85 at that time, and I elected to store it — and finally got $4.40 for it. They said if you store it you can almost always gain, but I sure didn't.

We've got about $20,000 worth of machinery and equipment. Not much compared to some. I supposed I could get a million and a half for the land; and I'm about $160,000 in debt. This year I just about broke even on costs and income.

Trying to figure out how to divide the ranch between all the children is a problem. We've made out wills. We've thought about incorporating, but there wouldn't be any tax advantage. Probably my boys will just take things over and operate. I might just lease the ranch to them. We have an insurance plan with some estate planning in that — enough to pay transfer taxes anyway.

The only money I ever borrowed was to buy more property or for operating expenses. I just make small improvements as I go along. My attitude toward taking financial risks in general is to go slow. Out here on the plains, people are buying land and leveling it, making 10 or 12 foot cuts. Now that is expensive, $500 an acre maybe just for leveling. They borrow the money. It scares me just to look at all that equipment some row-crop farmers have got. Some of them have ten acres of new equipment. A harvester alone is $60,000. Well, I bought a harvester in 1947 for $4,200. And I was selling grain then for $3.75. Recently I sold wheat for $4.40. At that income, I can't pay for a new machine. I raise wheat every year, only not on the same ground. I farm a piece of land over three years. I dryland wheat one year, and then let it rest for two. So I don't make much money on my ground.

I have been using my old machinery, keeping it going. I still use my 1947 combine. I'm unusual, though! I never ever bought a new tractor. The first tractor that I bought was a Cat '35 diesel and that was in 1936. I still keep that going. When it breaks down I hire somebody to fix it. It is getting kind of hard to get parts for that one. I had an HD-7, which is army surplus, for 15 years, but now I have a Cat D-7 that I have had for about 13 years. I have always bought most of my machinery secondhand, except haying equipment. I've had two new balers, and hay rakes and hay loaders. It is tempting sometimes to buy equipment, but it is so expensive. A new baler would be $15,000, and a swather would be $20,000. At today's grain prices, I just can't swing it.

Now my sons say, "Why don't you buy a new combine? Dad, why don't you buy a truck? You spend more for hauling than you would paying for a truck." I say I don't. An 18-foot truck with cattle racks and everything I need on it would be $25,000. And my hauling don't come to more than $1,500 a year. A friend of mine did some enterprise accounting and found his truck wasn't paying. Yet you can't rent a truck when you need it, because they are all busy. A person in that situation is really stuck. You have to buy the truck, even though it doesn't pay for itself.

Take grain. When we used to put it in bags, it would be banked out — somebody would haul the bags from the field and put them in a pile. From there they would be hauled to the warehouse. Now grain is handled in bulk. The truckers just bring double trailers and leave them, and you dump the grain in those. The grain in the field goes over from the harvester into a bankout wagon, and from there to trailers. You usually sell it at the roadside and then they haul it wherever they want to, usually Stockton or Sacramento.

Of course, the growth of all these different kinds of services now that we contract for, like truckers for harvesting, means I am less independent than my dad was. I have more business with other people than my father ever did. Maybe it's good for the economy as a whole, but it is kind of hard on the farmer's pocketbook. It makes farming easier physically, but not financially. When my father was farming, he wasn't out these expenses. And then he didn't have TV and electric lights. We had a coal-oil lamp until we got the acetylene lights. It is harder to make ends meet as a farmer now — mainly because we want more services, we want more things.

My sons want to keep the ranch going, but like the rest of their generation, they want money now. It is pretty hard for kids to stay in farming when their friends are working regular at the service station for $4.50 an hour, and they're not getting that at the ranch. My son-in-law was telling about some kids getting $9.00 an hour working up at the canal, placing rocks. Of course, the boys want that too. They forget that that job only lasts for maybe a month.

Along about 1930 there were more farms than in all the history of the United States, but since then the number of farms has been going down steadily, year by year. I think it is a manipulation. The Committee for Economic Development came up with a plan to remove so many million people from farms in five years. They did it, and are continuing to do it, economically by suppressing prices. And the price of land keeps going up. Farmers say, "Gee, if I can get that price, that is more than I can make on this ranch in a lifetime." I look at land like this: It is not a replaceable resource. This country is going to be like it was in the Old Country (if I remember my history right). The way things are going, land will belong to the exceedingly wealthy, and they will lease it back to us peasants. But I don't think these wealthy investors, even though they hire somebody to take care of the land, are efficient. You can't just jump into agriculture and then jump out. If it looks good one year, they get into it, and the next bad year they get out — but family farmers don't get in and get out. They are more committed to the land. But they have to be willing to make sacrifices. They can't live like uptown.

HELEN O'LEARY, AGE 62

My mother and dad came here from Missouri in 1937. My grandfather had died and left his place, and my father thought he could farm it, but he found out there wasn't enough there to make a living. My father became a railroad man, and that kept us going here and there. Then he was injured very badly and they came out here. They happened to settle in Colusa County because it was a farming area. Dad just did ranch work. When he couldn't do that anymore, he moved to Plainview and became a night watchman until he couldn't work anymore.

I was born in 1918 in Idaho. With the railroad Dad was always being transferred. It was hard for us to keep going to school. Then I had mastoid trouble, and my junior year was as far as I went in school. I got married young, in Missouri. I came out to California after my parents did, with my husband and three children in 1940. But there were family problems, and later we got a divorce. My mother and father worked up at a ranch in Elk Valley, and I went up there too. I drove the school bus for the little Wilsonville school and picked up all the children along the way. The little old man on the ranch was like an uncle to me. My parents worked for him, and I stayed there too, taking care of this old man, Uncle Frank, at $60 a month. The place was three miles from O'Leary's — Pat was ranching down below us.

Now Pat had been a friend of Uncle Frank's family for a long time. Pat was renting the ranch from his mother, who lived down here. He had already been married and divorced. He and I met through the Farm Bureau meetings. There was very little to do in that valley. Once in a while we would have a dance at the community hall. Pat's little girl Katie was maybe three or four. His sister was living there, trying to keep house for him, but I could just see that Katie needed me and I needed her.

Pat tried to keep Katie as much as he could, but she would stay out here sometimes with her grandmother. I always thought she was like a potted plant: She would go wherever it was convenient for her to be, so the family could go on with the ranching. When Pat was working, he couldn't have her there.

I could see so many things that could be improved at the O'Learys' and I knew how to do them. My mother's family always had a garden. They didn't buy anything, only went to town to get supplies. It was a very self-sufficient way of life.

My grandmother was a midwife. She never went to school, but she brought many, many children into the world and never lost a case. At two o'clock in the morning or anytime, the old phone would ring and Grandma would get up and saddle her horse, Snow, with a side saddle and go deliver these babies. When she and Grampa were first married, she worked for 25 cents an hour laundrying, while carrying a baby on one hip and one in her womb. She raised her own geese. And every time one of her girls would be married, a gift from Grandma was two pillows and a feather bed. She canned. They did their own butchering in their smokehouse. She made her own soap. She had an ash hopper and carried all the ashes to the outside, poured water over them, and then caught the lye in a trough. Then she would take the fat from butchering and mix it with the lye to make soap.

So I came from a very hard-working background. This is very touching to Pat. Sometimes he says, "I don't know how I would have ever managed without you, Mom." But when we got married, I could see so many improvements that were needed around that place that no one else did. I said, "Boy, what I can't do!"

I came to town about the time Uncle Frank passed away. He used to say, "Helen, you and little girls have to have a place to live." So he had bought a house and six lots for us. We lived there and so I had some property of my own. This was in the late 1940s or so.

Pat and I kept company for four years. Then in 1952 we got married and I went back to the Elk Valley ranch to live and rented out the house in town. I had cleaned up the empty lots and built more houses, to help support me and my three girls. When we got married we had my three daughters and Pat's one daughter together. And then we had the two boys in 1954 and 1956. By that time the girls were all in school. The oldest girl was 15 when Joe was born, just starting high school. A bus could have taken them to high school in town, but it was a 100-mile round trip, so we made other arrangements. I had the girls all boarded at the Catholic Academy in Round River instead. That wasn't good — we should have had them in public school because they missed out on a lot of things. The girls will never forgive us for it. But they all four graduated from the Academy.

The little country school in the valley closed in 1964. There were only seven children in it then, together in the one schoolroom, first through eighth grades. Dick went to school there one year. But when he got to be a second grader we decided to bring the boys to Plainview so that they could go to school here. Dick was a little immature. He didn't want to go to school. At Christmas the teacher said, "Pat and Helen, I think you should keep Dick out for another year." Well, it was so difficult for us to get him to school anyway, that we thought if we held him out we would never get him back there. So we brought him out to Plainview, and he stayed with my mother and dad. That was a sad thing, because it just broke his heart, he was so homesick. I finally said to Pat, "It is about time we did something else." We didn't have a home here in town then. I came out and scouted around and found a piece of property and bought it, and we moved. By that time, the girls weren't home any more.

During that whole period of ten years while we were on the ranch, I was out buying property. I started from the one house and six lots that Uncle Frank had given me, and got up to 15 houses and lots. While the children went to school, I would paint and paper and get the rental units fixed up. That was when you could buy a house and three lots for $4,000. And building was cheap; I built two rentals for $1,200. I have a pretty good business sense — if it is there, I can see it. I have been in this buying and selling a long time. We had a big old Victorian home here in town first, but we bought this place outside town because we wanted a little acreage. We have a lot of fog here, and Dick was nicked by a car on his little bike going to feed his 4-H calf one morning. I wanted to get a place where the boys could have their own steers so they wouldn't have to go down that highway. This place had three acres and a barn. The price was high, but I thought I could stand to be in debt much longer than I could stand to be without those boys. I went to the bank and borrowed the money, and in six years I paid for it. The rentals paid the difference. I'll say this: I had to do without many, many things while we were at the ranch. My work was very laborious. With this rental business that I have built up, I have made my work easier — and without bothering the ranch budget.

Pat's father carried a real heavy stick. They came up the hard way: "When I say, jump, you jump!" You didn't dare to ask for anything. When our home burnt, we were still pumping water in the old tank house up there. I don't know what kept us all from dying of malaria. I couldn't find out why the water wasn't coming through the faucet, so I crawled up on the tank house one day to find out what was wrong. Well, the water was terrible, because the cover had blown off I don't know how long before. It was smelly; we pulled a dead snake out, and it was full of leaves. And here we were, pulling this water. Oh dear, I just begged for a pressure system. I'm sure in all people you put too much emphasis on one thing and something else suffers. Right? Well, I'm sure we could have installed a pressure tank and pump for $500, but Grandma wouldn't do it. Pat wouldn't put it in himself, because he didn't own the ranch then. I look back now and I say, "Boy, if I'd had the fire in me than that I have now, I would have bought it and put it in myself." But she owned the house, even though Pat and I had built the biggest part of it out of our own pockets. You put up with things when it is family you are dealing with. Anyway, I said, "Pat, never again do I try to save $500 just to save it." They wouldn't spend $500 to save $5,000, is what I am trying to say. Even today the boys will say, "Oh, Dad, you poke along all day long in that field with that one old engine, when we could get a new tractor and get it done and do something else." But that is the way he grew up. (On the other hand, that old tractor is all paid for.)

Now I think sometimes that Elk Valley wasn't very kind to Pat. His mother and dad really depended on him, and Pat was just the kind of person who loved the ranch, and he wouldn't let them down. His brother didn't have the same get up and go. I mean, maybe 10 o'clock would still catch him in bed. If Pat was working on one end of the ranch and Tom someplace else, well, if Tom broke down, he would just wait until Pat came. He just didn't have the initiative. But he was a brain — one year he went to the University of Washington. But his life has been kind of sad. He has done very little ranching, really — he just sort of went through everything. Habits can be very expensive. One thing about families, it is not just an economic thing at all; you are dealing with people, and there are a lot of hard feelings sometimes.

When we left Elk Valley we were sad about leaving the ranch. The little boys thought that I was losing my mind. They wanted to stay. And we had to tear up that home. Ripped the carpets out, and tore the drapes down. We just gutted it. We had worked hard to have all that, and all those years I never knew what it was like to have wall-to-wall carpeting. We got to use it for such a short time. The girls and I had hauled the bricks and cleaned them to build our fireplace. Then I had taken the old staircase and the newell posts out of the old Clinton place and I refinished every spool in that, and put it in our own home. That is another thing that hurt; I had thought it was really going to be my home, and I fixed it up so pretty. All the good things I put in there — it was jealousy in the family that caused us to lose it. I said, "The first home I had, some other woman wanted it. The next home I had burned down. The third one, one member of the family wanted to force a sale." You know how it is. When money is involved, they are all starving to death.

It happened because my mother-in-law died without a will. Papa had made a will, but Mom never signed it. The will is why Pat had put himself into that ranch; the boys were to get the ranch and the girls to get the money. But Tom was a problem, because drinking was part of him. There were only a very few acres in that whole place to farm. Pat was trying to make a living for us. We had to get something out of it. Pat had rented that ranch just to support his mother; for every dollar that Pat took off the ranch, 50 cents of it went to her. Even our accountant said, "Pat, no one in the world would rent a ranch like that."

Mom didn't sign the will because there was no money for the girls. Bless her heart, she had no alternative. If she wanted to divide the inheritance equally, the ranch had to be sold. On her death, when the seven brothers and sisters wanted to divide the estate, they were going to sell the ranch. Eventually it was sold to Pat. We already had the Clinton place, which Pat owned before we were married. After that, in 1954, we bought the Janick place. So when he bought the homeplace, he already had the other two. But the homeplace was important to him. It was his life. And it was right in the center of the other operations, our corrals and all.

We didn't know if we could get the money. And we didn't know what Mary, Pat's oldest sister, was really going to do. She was the one who wanted to force the sale. We went to an attorney and he told Pat to try to talk his sister out of it. A public sale would be the saddest thing the family could do. Mary wanted a sale because she thought there might be more money. We thought that wasn't fair because we had put our own money into the house. The attorney said, "Well, that is bad." He said, "Do you think you can get the money to buy her out?" Pat said, "I don't know. I haven't been trying, because if she goes along with this sale, there is no use anyway." He knew that someone would come in and buy it. We had already made up our minds that if the homeplace was going to sell, we would sell the other property and move. We had already gone to Canada to look at a ranch. The attorney told Pat, "You can take everything out of that house as long as it doesn't do any structural damage." So we took the built-in units out, the stove, the light fixtures, the drapes, the carpeting, in preparation for a sale. All that work, putting it in, and all that work taking it up!

Pat and I went to the Land Bank. We could bring in the money. Pat had to pay each one of them off. He bought all the cattle and the whole ranch. It was a real thing to do. One time we were driving over to the Bartlett Spring, talking about what we should do. We still owed money for the Clinton place and the Janick place. I said, "Pat, if you buy that estate, it will kill you." And Pat said, "Mom, if I don't buy it, it will kill me." So I knew how he felt. I thought, well, let Pat buy it and die happy. It was his whole life, I knew then.

This is what made me feel that I should just get in there and dig deeper and harder. I'd see this and that property around, and I'd say to him, "That is a good buy." I begged him to buy other property. We could have bought a beautiful ranch for $600 an acre. But Pat would be afraid. He says now, "Mom, you were never afraid." He'd say then, "Mom, you are going to get a bad deal." But I am proud of myself — I have never had to ask Pat to back me up or bail me out at any time. Now I could sell out for quite a bit. The money I've made helped get some food on the table so the ranch income could do something else. And it helped the little sore spots with the family. The ranch hasn't been all that much of a money-maker. Things like sending Joe to Europe, helping the children get their pickups, and helping the girls get started when they got married — all that I was able to do because of my real estate.

I have a room out here in the back where I refinish furniture. I refinished all the furniture that we keep house with here — the whole dining room set, 13 pieces. The old homes I'd buy would be full of furniture and I'd refinish some of those old things. I lost a lot of pieces when our house burned, and I couldn't afford new furniture and I didn't like it anyway. So I refinished things from garage sales.

Then I've always done canning and jam making. I pick up peaches over in Sutter County. I used to say to Pat before we were married, "How could you pay this horrible grocery bill?" He never had anyone to help him. I made my own bread. I used to help Pat scrape the hogs. I picked the turkeys. We had chickens and our own eggs. We had milk cows. The girls did the milking, but I made the butter. I never had any hired help on the ranch. I used to have 14 kids that would come to the house. I always had a lot of people sitting down to dinner. One time in summer — I remember it was 125 degrees in that house if it was 90 — I had an old gas stove with the warming oven where you bake, and no cooling. My friend said, "Helen, I don't know how you do it." I had 32 people that night for dinner and I baked berry pies.

I never did field work like driving the tractor, except for moving equipment from one ranch to the other. I had all I could say grace over in the house. I had to laugh at Pat. He would be dead when roundup time had come, with all that roping and marking and branding. But that was farming. And there was always mother at the house cooking for those cowboys. And that is a big job too. I would bake bread — boy, at one sitting, four loaves of bread would be gone.

I know all about ranch work. I helped Pat brand 99 cows myself about five years ago. The reason I know it was 99 is because I said, "Pat, couldn't you find one more, and make it an even 100?" Just Pat and I were here, I don't know where the boys happened to be. He was ready to brand, so he said, "We'll do it, Mom." So the two of us did it ourselves. We had one iron heating and one iron in use. I kept the fires going. Pat did the castrating. We put on a repellent to repel the flies, and I would run the paintbrush, and vaccinate; we have a gun that holds so many cc's. Pat would dehorn — I can't stand that. I would run the cattle in, and help him separate. We worked about 12 hours that day.

The two of us have worked many a time by ourselves. I have been on many a mile on horseback with Pat, and have driven cattle at 4 o'clock in the morning, when we wanted to get them down to the neighbors' scales to weigh them for the trucks to pick them up.

After we came out here to live, we would go back to the ranch and stay the three months up there during the summer. We never left the ranch entirely. I would still cook if he was going to have a crew up there. And I was the errand person, over the mountain to town for parts and then back for the meals. I saved Pat many a trip. But sometimes when I went to Colusa they'd ask me this and that about what I wanted, and I'd be stuck. Once I said, "The next time I need buttons, I am going to send Pat to town after them."

I did sewing. There weren't any curtains. They lived very humbly up there. Mama never knew what it was to have anything. They couldn't! And oh boy, would she be strict! Like this, for instance: "Grandma, why don't you let me wash your curtains, and wash and clean up here?" She would just fly right back and say, "Helen, don't let my house bother you." She was very snappy with that old Irish in her. It was really a way of survival. You had to be stern.

Between the time that Pat's dad and Mama died was about 13 years. During that time she left the ranch things up to Pat. We didn't ask any favors either. If I wanted to paint or clean up the mess, I knew she wouldn't give me permission. She would say, "Why don't you sit and talk to me? You just run all over." But she enjoyed the things I was doing, even though she never had flowers, or turkey coops, or anything. Once I said, "Grandma, I can't sit down because of that pump out there. You can't push water uphill unless you have something behind it." I was trying to fill my washing machine, so I excused myself and pumped it up again. She said, "Well, if you think I am going to buy you a pressure pump, no!" She wouldn't do it. And Pat was the same way then. I said to the boys, "You should have known your father 25 years ago. You wouldn't have gotten off so easy then."

The boys are getting minds of their own now. They want different things. It was like two families that we had. As Mother and Dad got older, we became more mellow. We haven't pushed the boys like we did the girls. The girls did a lot of work on the ranch. I just wish that Joe and Dick had the knowledge and the experience behind them that Lucy and Katie had at their age.

The boys really want to ranch, but they want Dad to retire. Pat is getting old, he is 68. He goes up there every day, but by 10 o'clock, he is all through, he is tired. He is not really doing anything up there, but he wants to know what is going on. He is so proud of his sons: Tom, his brother, is an old bachelor and will never marry, so Joe and Dick are carrying on the family name. It gets pretty touchy sometimes here, though. I keep telling the boys to just go along with Dad until he knows he can't do it anymore. But they want to take over and do things their way. Had he done what I felt was best — he had these two boys coming on — we could have had our irrigated pasture out in the valley and still have gone back to the mountains with our cattle. But Pat has always been real conservative. He was afraid. It was all around him. The boys get impatient and say, "Why didn't you buy . . .?" They look at $2,200 an acre land and don't know how they could make it. The boys want to do things and he will say, "No, you had better wait." They nicknamed him "doodlebug," behind his back.

Pat said a while ago to them, "I will pay you boys so much a month to stay up at the ranch, and you'll get board and room and gas." But they are not going to stay up there. There is nothing there for activity. They are 22 and 24 years old and social life is important. Joe said, "All right, Dad, what are we going to do in the evening when all the work is done?" Dick has a wit that won't stop, and he said, "Oh, I know what we can do, Joe — we can play Frisbie with cow chips!" They could come down to Plainview for the evenings, but then Dad would say, "I'm not going to pay your gas bill so you can run all over the country."

Recently Dick fixed an engine for Dad. And Pat is pretty critical in lots of ways; he won't say, "Dick, you were right." But that is what these kids are waiting for — he could do it, you know. Just yesterday one of Joe's friends came over, and they were going to the ranch to feed the cattle. Joe wanted to dehorn this old cow. Well, the weather was nasty and Dad said,

"You had better not, Joe — they are kind of wild anyway. If you don't make the first loop, she might give you a hard time." Joe decided against dehorning the cow, but he was still going to move the cattle to the other field — get the hay in the truck and let the cattle follow. Well, Joe didn't come home, and he didn't come home. So Pat said, "I know what happened. His truck got stuck and he didn't do this and he didn't do that." He had it all figured out in his mind: Joe went over to get the truck out, and he got his pickup stuck, and then he probably got the D-7 stuck. Pat had me walking the floor, because he was walking the floor. And because he was worried I thought, "Oh my goodness, what's happened to that kid?" I was calling as many people as I could up there to see if Joe had come out. They didn't know. But this is what had happened: Joe O'Leary moved his cattle, fed them, and didn't get stuck; it went beautifully. A friend from Plainview was going to get something somewhere, and *he* was the one who got stuck; and Joe had a heck of a time getting that Ted Higgins out. So Joe did a beautiful job on what Dad said he couldn't do. But Pat thinks if he is not there, it is not done right.

The boys want to ranch, but the thing right now that is so discouraging to them is Pat trying to keep everything under control. They say, "Mom, keep Dad at home." And I can remember years ago Pat would say, "Oh my God, there comes Papa." History repeats itself!

Dick is picking up a couple of classes in grasses right now at the college. Dick will go somewhere else if his Dad doesn't turn him loose. I have been trying to persuade Pat to retire. But he says, "The boys can't do it, they will fall flat on their faces." I say, "Let them — that is what it is all about. They will build themselves back; they have a lifetime to do it. You and I have about 20 years." He is just going to have to get used to the idea that they are grown up. The whole story in a nutshell is that Pat doesn't want to give up. But he is going to lose his two boys. When he gives up, they might come back to ranching. But by then maybe it will be late for them to get a good start.

If our boys had real responsibility they would buckle down. I said, "Pat, why don't you tell the boys just what a debt we are up against?" Pat said, "Mom, I think they are too young to know." But they are not. Dad is very close. It would take a good Philadelphia lawyer to get anything out of Pat O'Leary. Our boys are very young at their age because they have been protected. But I think the time is coming, if it were put upon them, they could do it. Pat says, "They will tell all their friends how much we are in debt." I say, "So what? It's no disgrace." But this is the way Pat feels.

## JOE O'LEARY, AGE 22

I was born in 1956. Farming is all I have ever done, so it's all I really know. I grew up working cattle, cutting grain, planting grain, feeding, fixing fence, anything that needed to be done. I started out pretty early helping Dad. I didn't care much for school. I graduated from high school in 1975.

After that I worked for a construction outfit, building mostly pole barns, shops, and houses. I worked in rice, driving bankout. And I have worked in shops, run engines, cut grain, and lots of things around here locally. Nothing else interested me really. I'd rather work than go to school. It wasn't the homework I didn't like. I just don't like listening to someone lecture all day. I don't like being cooped up.

I did go to Europe. I got a deal in the mail one day, a people-to-people ambassador thing. Mom and Dad talked me into going, so I went during the summer of my junior/senior year. Mostly we stayed in hotels, but we stayed in a convent one time, a private school. We stayed with families too, in London, Germany, and Holland. Mostly it was just travel, for two months. I like to go different places and see different things. I went to Alaska twice, on vacation with Mom and Dad and my brother, and we went to Canada four or five times.

Right now I am buying cattle to put a gain on them and then get rid of them. When we have enough money we'll go into a cow-calf operation. My brother and I want to get on our own. We have a few of our own now, but not as many as we want. We need some summer grazing ground. We tend to run out of feed on the ranch. It is hard to say what we'll do. I would like to make as much money as we can, and get enough cattle so we won't have to raise anything else. Maybe get some ground out here for a little wheat or some corn or beans. But that is way down the road yet. It depends on how cattle prices go. If cattle prices fall tomorrow, it would set us back a year. I like ranching and raising cattle. You don't have to put up with all these people out here. I like being in an open area. That is why I never go to the city. I just can't stand all the people down there in Sacramento.

I'd like to change a lot of things about the ranch. Just about everything, in fact. But until the end of the government that I am under there is no way to make any changes. If it was up to me, I would get some equipment that I could do something with. My dad has old equipment — ancient! It is hard to do anything. A two-hour job takes all day. And we don't have the right tools to be efficient. I know quite a bit about machinery, I can fix things, unless they are really something complicated. Anything he has out there, I can fix it, but you get me out here with hydraulics and things like that, I don't know that stuff. But we don't have anything to work on the machinery with. The only tools we have are what I have. And I have only enough tools for me.

Dad doesn't see where he could be saving time. Not so much in labor but in working days. You could go do something else. Let's say your engine breaks down, and it is going to rain in a couple of weeks, and you know your ground has to be covered in time. If we had the right tools, we could have the engine fixed in two or three days. But up there, fixing it takes us a week, so we are losing money.

The cattle operation is all right. But that gets back to not having the proper tools too. It could be better. We did some leveling up there. In the valley you got all these low spots and high spots. When it rains, all the water builds up in the low spots and drowns out all of whatever you planted. It would be possible to level some more land up there, but we don't have anything to do it with. We've got one scraper and a Cat-7, which is shot. It needs a motor put in it, which probably will never get done. I guess Dad thinks he can't afford it.

Jim and I would like to be in partnership. At the moment, we are still working for our dad. We have some of the cattle up there and Dad has some. We just hope that cattle prices will stay up. I've thought about doing a few other things. I might be a mechanic, I enjoy that; or be a contractor, I was doing all right at that. When I was putting up pole barns for Don Sands in Plainview I made $1,000 in ten days, and three of them weren't full days, about four or five hours. It was easy money, though it is pretty hard work. We had to walk on trusses high off the ground — the highest building was about 45 foot. We were walking on an inch-and-a-half truss with a four-inch gap in between them. I didn't think it was too bad. If someone would be down on the ground and yell, "You are going to fall," as soon as you think about it, you kind of get a little uneasy. But other than that, you know, all you have to do is watch where you are going to put your next step.

There is no way I could keep that up with cattle too. You can't do both. When a construction man quits for the weekend, the barn is not going to get sick, and if his building goes down, it's not going to be crippled. Diseases and things like that just don't take a holiday. With the cattle, you have to be with it every day. At least you have to check them everyday. You don't just say, "Oh, here is the grass — go eat it." You have to go out there and see the sick ones, and anything that might turn up. They are not going to stay in one spot. They get out and get around. If they get in the road, they may turn up 20 miles away, wherever they want to go. Most of our cattle are on the hills, with a few down in the valley. Some of them we can check out with a pickup, and some of them we have to go on horseback.

My brother and I work together all right. We could work for other ranchers right now. But we don't want to stay just a few head here and a few head there. We want to get a bigger herd. There is some ground we are looking at now, 500 or 600 acres, that we could put some wheat on. We could rent the machinery we need. Between the two of us, just on our place, we could probably handle about 600 brood cows. It would depend on how long we wanted to keep the weaners, and how many heifers we would keep every year, and how many old cows we would sell. On better years we would keep more. If we didn't get enough rain, we couldn't keep as many.

But my dad will never retire. I don't know whether it is that he likes what he is doing so well, or just habit, but he will never retire. He will just go on working.

Harvesting almonds with mallets and poles into canvas catching frames was laborious work before World War II. Reprinted with permission of the San Joaquin County Historical Museum.

# THE NEW BREED
## The Breidenbach Family

The older Breidenbachs live in a simple stucco house separated from their son's residence by a large plowed field between. A few old almond trees and grapevines surround the house. In back are large storage sheds housing tractors and combines. Wilhelm is pale and very old, nearly deaf, and it is difficult for him to stir himself out of the reveries of age to talk to a stranger about farming in the long ago. Martha, eight years younger, always attentive to him, talks much more freely. Her face has the tranquility of a woman whose life has been contented, whose relationships have been happy.

Fred, their 54-year-old son, is a muscular, vigorous redhead who would rather be out doing than talking into a tape recorder. His look and manner inspire confidence. He might be Thomas Jefferson's independent yeoman — guardian of the nation's values. Fred's wife Loretta is also active, energetic, proud of her family and of herself. Her quick smile and outgoing nature seem to preclude any thought of the darker side of life.

WILHELM BREIDENBACH, AGE 86

I was born in Yolo County in 1892. My folks were farmers. I worked for my father on the ranch; then my brother Pete and I came up here to rent a ranch together. It was owned by the Hershey family and was part of a big estate. They had lots of land and had it rented out to different people. We came from a big family, and when Pete wanted to get married he couldn't stay home, because my father was still farming and had other children. Pete and I had to branch out.

I went into the service for World War I, but I didn't see any action. I didn't go overseas — I would have gone, but just at that time the Armistice was signed and we never got across. In fact, we got on the boat and sailed past the Statue of Liberty. Then they stopped us and we turned around and never got any further.

We farmed different then than now. With mule teams, and no irrigation. Everything was dry. Every other year we raised grain and then summer-fallowed in between. We raised barley, mostly. A few years we had sheep too.

My brother and I rented this 1,000-acre ranch quite some time. Then he left to go rent somewhere else and I farmed alone. Did we have hard times? I don't know what you call hard times. We made a living, it was all right. It wasn't like now.

## MARTHA ZELLER BREIDENBACH, AGE 78

I was born in 1900 in Campbell. I became a schoolteacher at the Thorn Creek School near here, close to where Wilhelm was farming, and that's how we met. I taught a year and a half and then got married in 1919. My folks farmed too. We had a big family — ten of us. There were about ten in his family, too. Our folks all came from Germany. My father came and worked and bought some land here. His folks were the same way. Our parents spoke German to us all the time when we were youngsters, but when we went to school we lost all that accent. When we got to school, why, when we came home, naturally we talked English with our parents, and they kind of forgot their German.

We boarded the men when we had harvest help. We had a cabin for them and I had to cook three meals a day. About six men stayed the whole time. The harvest lasted roughly 12 days. But if we had north winds, then they would stay as many as 15. When we put in the grain, Wilhelm had help too. Usually we had two men then — one would sow and one would drive the tractor.

Half the ranch would be in grain, and half was summer fallow at any given time. We didn't fertilize like we do now. You had to let the land rest one or even two years between crops. Of course, this land around here isn't the best, like it is toward the river. We grew barley. Now they plant wheat more. They have a different kind now than when we planted. We were always afraid of wind, which would shatter the wheat. This wheat now can withstand the wind better. Barley was used for feeding animals and for milling too, for flour. People around here didn't have the production like they do now. It wasn't a big income. But gas prices and all weren't like they are now, either. Everything was on a smaller scale. We didn't get rich, but we got by. We stayed home, we had our friends. We had more company at home. We would go visiting with the family. Now, of course, everybody eats out and goes to Sacramento for shows. But we didn't have baby-sitters.

We have been here on this place since about 1935. Wilhelm rented land for about 20 years and then in the 1930s we were able to buy some. First we had about 600 acres, and then we bought more later, from a widow whose husband had just passed away. You know it is hard, not having the man to do the work, and she didn't have children who wanted to farm. She liked town anyway, so she sold the ranch. Then we built this house and have been here ever since. We bought an orchard later.

We had a milk cow, but no cattle. I planted a little garden, but the soil wasn't so good here. I guess I could have fertilized — they did sell that in the store — but I never gardened much because I thought it was awfully hot. With a garden you have to work at it. And I had men to cook for, and not much time, with three sons. I never worked outside, I had plenty to do inside.

The Farm Bureau in those days had meetings once a month in the schoolhouse. Wilhelm liked that, but the lodges, they were not for him. The Farm Bureau was more active then than it is now. The community really would go. The head of the Farm Bureau would come down and tell them all what was going on. After the speaker they would have a little entertainment — someone would play the piano and sing, or do something. And they always had community singing at the beginning to open it up. My husband enjoyed that, and the youngsters did too. We would always take our family. About 15 or 20 families would go. But the Farm Bureau doesn't have those community meetings anymore. It must have been 40 or 50 years ago, the times I'm talking about. They stopped here about the war years.

There never was a clear time that we retired. Wilhelm got to be about 66 and the three boys wanted to farm, so he let them take over, but he still helped them; he was always out there. We encouraged our sons to stay in farming, we hoped that they would. They didn't go on to school. They liked farming, that was their life. After they were in the service, when they came home, Wilhelm took them in as partners at first. Then he just dropped out and they kept on together. They changed some things: They started fertilizing, they put down wells. This is scrabbly land around here, although not what you would call poor land. With water, it can raise row crops. The boys rented land and now they have branched out and bought new land on their own.

They are busier all the time than we were. When we farmed we just put the grain in. Now they plant beets; they have almonds, they irrigate with sprinklers. When we had the barley in, we had to prepare the summer fallow in the spring — the 500 or 600 acres — for the fall crop. But we had leisure time in between. We weren't always working on our land. In a way our sons work more than we did. But then, of course, they make a lot more money. They are always busy, putting in grain and fertilizing, spraying, moving sprinklers in the orchard. My husband did all of his own work except at planting time and harvest. At slow times he would be patching sacks for harvest and getting ready. Now everything is bulk — they don't do the hard work like he did. They spray, which we didn't do. If the bugs got in, well, we didn't know about it, I guess.

Farming is a little easier now. At least you are pretty sure you will get a crop every year. They are sure, we weren't. We depended on the weather, now they don't. We had to have rain. If they don't have rain, they can irrigate. They have more machinery, but things are more expensive. We would get a tractor for $5,000 and that was a lot of money. Well, tractors now are about $75,000.

We have been farming in the area for 57 years. Our neighbors are pretty much the same families. Some of them have passed away and if their heirs don't want the land, they sell it. I have some friends who sold their farm. They had no children and the husband was getting along in years, so they moved to town. He has passed away now. She sold all the land. She said, "I don't want to bother with it." It is a worry. Older people, things worry them more.

## FRED BREIDENBACH, AGE 54

I was born in 1925 and grew up on this ranch. I only had grammar school and high school, right around here. I've never had any formal training in agriculture. I just learned by doing, and through my father, and by trial and error. I went into the service in 1943, the year after I graduated from high school. I spent two and a half years in the service. Went in as an Air Force cadet, went through pilot training, and ended up as a bombadier. I never saw any active combat, though. The war was over before I was scheduled to go out. When I came home I spent two years farming a little with my father and did a lot of hunting and fishing, which I love. Then we formed the partnership with my brother in 1947.

This was a dryland farming operation through 1946. Then we started developing some of the property, leveling land and drilling irrigation wells. We acquired an almond orchard in the early 1950s. From then on we've continued developing our farming until the present. Three-fourths of our farming now is based on irrigation wells. We hope to get water from the new canal. We are developing land to be prepared when the water is available.

Formerly we had livestock — sheep and some hogs. During the dryland farming time many people in the area had stock. They grew their own grain and fed it; it was a way to make a few extra dollars. But since irrigation, livestock has gone out of this area. The fences have all been removed, and we have leveled most of the land and farmed it all. Livestock interfered with row crops in many ways. During the winter months, to have a place to run the stock, we had to leave the ground idle — and that's something that we don't do anymore. We never leave fallow fields. We farm every acre every year. We've had a steady progression toward row crops like sugar beets, and away from grains. The income is much greater from beets than from grain. We still grow wheat, some barley, some milo. From year to year we juggle the crops. This year wheat would be around 650 acres, the barley 100, and the milo 100.

How do we decide from year to year how much we are going to plant? Well, some land is planted to grain every year, because it is not irrigated. Price is a factor — that and the year of the cropping pattern that we use. If we have a field in sugar beets, our normal practice is to winter it over and harvest in the spring. If we get the beets harvested in April, we go in with milo. If the beets don't come out early enough, we could very well let the field lie idle until the fall.

We sell on the open market, so we consider the market when we make our decision to plant. Sometimes we sell the grain at harvest, or we may hold it through until the end of the year. Many grain associations have done well. But when you sell your grain independently, you get your money right away. Most of the associations pay over a period of six months. From a farmer's point of view, there is more flexibility in marketing the grain yourself.

We have 450 acres of almonds now. Harvesting 25 years ago was done with a sled and a couple of almond sheets, 16 by 32 feet. We would drag the sled through the orchard with a mule or a small-wheel tractor, pull up to each individual tree, and position the canvas sheets underneath. Then we would actually climb the tree and knock the nuts off by hitting the branches with a club, or take a pole to the little branches on top. The nuts were knocked onto the canvas holders, gathered by hand, and then dumped into the sled. After 10 or 20 trees, the sled would get full of nuts (hopefully), and then they were shoveled into sacks. Then they would be picked up by a truck or trailer and transported to a shed. We would fill the sacks only maybe two-thirds full; we wouldn't sew them, the tops would be left open. The almonds would be quite green and wet and would need drying. After running them through our huller we would sack them up again, this time sewing the sacks together. We put them on an air-dryer to dry the meats and shells down so they could be delivered to the Almond Growers Exchange. All of the hulling and drying was done on the farm. We did no shelling until 1960. Then we started actually cracking the nuts, so that we delivered nutmeats to the association.

We had many more field workers at that time. In the 1940s we would have eight in the fields, and possibly three or four ladies working on the huller to separate the hulls from the almonds. So there were 12 to 15 working on the harvest, typically, for us. We had braceros at that time. They would do all the knocking. The hulling or things that required machinery would be done by my boys, or some local help. In the shed we had local women — neighbors, my wife. The women didn't work out in the orchard.

Now all that has changed; it is all mechanized. We prepare for almond harvest with landplanes and equipment to get the soil perfectly smooth. We sprinkle the soil to firm it up so that it will be compacted and get a hard crust on the ground underneath the trees. Then we go through with a shaker, a machine that shakes the nuts from the trees onto the ground. A sweeper pushes them into windrows, from underneath the trees into the center of the rows, another machine — a mechanical device with fingers that flip the nuts up — picks them up, and a fan on the machine blows out the dust and leaves. The nuts are put into a trailer, then towed to the huller at ranch headquarters. We still separate the hulls from the almonds as we did years ago; of course the equipment has all been updated. We deliver the almonds in shell to the warehouse. We gave up shelling because we weren't able to do as good a job as the association. We were getting more chipped and broken meats than they wanted, so they were penalizing us. In our new equipment, "airlegs" remove the hulls from the nuts, so it is pretty near all mechanically done. One or two helpers work part time on certain varieties. Many growers don't use any help at all. When we shake the nuts off the tree now, we leave them on the ground until the hulls are dry, so we don't require drying any more.

We have about half the people we needed before; but many more dollars are invested in equipment. The thing is, machines don't get tired, and people do. And almond harvesting was very laborious work. We also don't have to worry about illness in hot weather, and doctor's appointments, and people being people, wanting days off. If we can do it mechanically, we can get more done. It is much, much faster and that is the biggest advantage of all.

Most of the almond harvesting now is done by myself, my two sons, and my wife's help when we need it. Out in the field we still have migrant workers to help us, but last year we had only four. Compared with 30 years ago, we leave less of the crop in the field. Knocking and poling was real slow, and it created problems in the huller. Tree limbs were broken, and we had more sticks and leaves and trash to contend with.

We also have better equipment for spraying. We control diseases better, and that eliminates a lot of problems. With a healthy tree, things just go better. Almond growing is much better than it was years ago. Production has increased substantially, not just per acre, but in overall acreage too. The almond industry in California has expanded because of these technological advances. The industry would have stayed relatively small otherwise. Almond harvesting the old way was really hard work — hitting the trees with a club took a lot of strength! People originally thought mechanical shaking would harm the tree, but it doesn't. It doesn't affect the root system. Even though you make think that the whole tree is shaking, the vibration is minor. You may get a little damage on a limb once in a while, but not nearly as much as when it was done by hand.

In sugar beets our operation has expanded too. We plant a field much faster than in 1947. We need less hand labor since we have mechanical thinners to thin the beets. We still use some hand labor for hoeing, usually one time through the field to get the weeds that the sprays missed. Thirty years ago the crew would have had to go at least twice through the field, and it was much more expensive. When everything was done by hand, the time element was really a problem. Speed of operation in general has its advantages because it minimizes the risk that something bad might happen. With a machine you can work night and day, which we do many times. Weather is a big factor. If we have just a few days of good weather now, we can harvest a crop, whereas 30 years ago it took maybe weeks to do the same thing.

The margin of profit in an average year, I would say, would be equal to years ago, but every crop is different. We've had real poor years on all crops, and we've had good years. We are in agriculture because we are making money, or else we wouldn't be here. But it is necessary for us to make our land produce more per acre than ever before, to pay for our machinery and our fertilizers and everything else. Labor is expensive, harvesting is expensive, fuel is expensive, everything adds up. Taxes are a big item now. Spray is used for everything, and much of it is done by airplane. Equipment is much more expensive, and we have much more invested in land. Production costs have gone steadily upward. We have to produce more because we have to cover these costs. You can't be an average or below-average farmer, or you won't stay in business. You have to produce as much as possible from a given acreage.

My two sons and I run this farming operation full time — my wife too! We have no year-round employees, but we hire additional help for the irrigation months and almond harvesting. We don't dig sugar beets ourselves, it is done by a contractor. We just grow them. We grow and harvest the 650 acres of wheat all ourselves. Bulk handling of grain — which really started about 1946 — has been very significant. We were real slow in California. Washington State was bulk handling its grain five years before we were. In certain areas hauling those sacks up to the warehouse, 100 pounds at a time, was really a tedious operation.

Our contractor gets labor for the sugar beets. He has a crew of 20 or 30 come in for a couple of days for hoeing. Sometimes they are winos or somebody who can't get a job anywhere else. They're not good workers because they aren't skilled, and sometimes they're sick. Some of them, when they get to the ranch, can't even leave the bus. A few of them will be good and they'll just go ahead and work ahead of everyone else in the field. Sometimes it seems that agriculture is expected to employ people who can't find a job anywhere else. And we have to train them. On these labor contracts they bring all kinds of people, and you see them getting a little bit worse here in later years. A lot of wives of Mexican workers do hoeing work, though, and they are good workers.

I'm a farmer because I like the independence. I can take a day off if I want, or I can work Saturday and Sunday and all night. My father always felt that way. If he wanted to take a week off and go deer hunting, he could. He didn't have to answer to anybody else.

That tradition of individualism makes farmers hard to organize. But education has changed farmers, too. They are more interested in group cooperation. And the study of economics has changed. Both of our sons went four years to college. They are interested in what goes on nationally, because that is what they have been taught. Maybe as this generation of farmers spreads its wings a little bit, it might be more likely to organize effectively. Of course communication was a big problem years ago. Now it's nothing to jump in your pickup and drive to Oakfield to go to a meeting. In those days it was unheard of, unless it was really important — you just didn't go. It was 25 miles! That's nothing now. But at that time you went maybe once a week, and everything that you had to do, you did it that day. Now we might even go down twice in a day.

I anticipate farming in partnership with my two sons until I retire, which is still a few years away. I was in partnership with my father and brother until we dissolved that. The coming of the canal would have forced a split anyway, but my brother and I both had boys wanting to get into agriculture, and they wanted to make some of the decisions. It was getting to be just too large an operation, too many managers. It costs us more money to farm individually, though, because we all had to buy our own equipment. It would have been more economical to farm it as one unit.

## LORETTA MACPHERSON BREIDENBACH, AGE 50

In 1928, when I was born, my father owned a cattle and wheat ranch in eastern Washington. The Columbia Basin then was cattle and grain and dryland farming, somewhat similar to here. Only it was larger scale, because my dad had about 5,000 acres of wheat. He planted half each year and summer-fallowed half. He had shorthorn cattle, all registered, which he sold to breeders in Oregon, Washington, and Idaho, and also to 4-H kids. He could always get $100 a head for each calf he sold.

I had two brothers and a sister. I was the second child. We were children during the Depression, but we didn't know those were bad years because that's just the way it was; everybody was in the same boat. It was very rural. My dad had come there in 1904 and his family started the town of Sandy Lake. There were two aunts and two uncles who had a store and a meat market and a slaughterhouse and a little hotel, and that started the town, along with the post office. It was a very family town! And they were all Scotch people.

We lived on the ranch. I always said I was raised on a horse. We had to get the cows in, and we supplied our own milk, meat, vegetables, everything like that. We heated most of the time by wood and coal. Years later we had an oilstove heater. They didn't have electricity until 1947, after the war. Not until I had gone away did it come in. We had kerosene lanterns, gas lamps. My sister had allergies, so she did most of the housework. I was always outdoors, because dad needed extra help, which he couldn't get during the war. So I drove tractor, I drove truck, and did anything that was needed. I was like another one of the boys.

I went to Washington State. I was going to be a physical education teacher or a home economics teacher. But then my cousin from Hardin here asked me to go with her on a trip, so I did. When we came back the semester hadn't started, and she wanted me to stay with her since she was an only child living here, so I did.

I was working in the bank in Hardin, and that's how I met Fred. We got married in 1949. I helped him farm from the beginning of our marriage. I had driven trucks for years up in Washington, so when I came here I drove bulk truck during the summers. I worked on the huller practically every year. When I couldn't take the kids, grandma would take them, and I would work a half day and she would work the other half. Sometimes we would feed the harvest crew, and we would take turns in that. The other sister-in-law worked too. Physically I have probably been more active on the ranch than the average wife. I still am — sometimes I'll load pipes, sometimes I'll drive a tractor, or help on the machinery. I like to work. I like the variety. I like to feel I'm helping.

In the last couple of years we've had a huge field garden in which I do most of the planting. We have four or five crops of corn. We start one crop early and when it is three or four weeks along, we plant another one. We have melons, a row a quarter-mile long. It's all for family use and friends. We had 70 tomato plants last year. You have a lot of friends when you have tomatoes! We give melons to the hired help so they can have fresh fruit. Fred and I do most of it. We just plant about six rows in one of the fields. I have my own little garden here too, and that takes daily care. I have peppers, summer squash, green beans, carrots, and beets. The field ground isn't hoed enough, isn't in good condition for root vegetables. It gets hard and we water only when there is irrigation water available, maybe only once every ten days. Here I have things like potatoes and cucumbers that have to be picked every day.

I go to Oakfield once a week and generally pick up some part or take a part down to be fixed. I do a lot of running back and forth. Bookkeeping I don't do — the boys do that because they took accounting. Sewing, canning, and cooking are all part of my life. When the Extension food technologist comes up here I take everything I can get a lesson in. I taught 4-H sewing for 12 years, Sunday School for 12 years.

I really haven't had time to join many organizations, because I spent all my years with my kids. I followed them everywhere! I just wish I had a dollar for every hour I spent waiting at the school. All four have been very athletic and very good in academics too, so they have been in everything — which is the way I wanted it. But being out in the country, I had to get them back and forth a lot — I went into town seven times one day, I remember! I used to take them swimming every day, too.

The kids have all worked on the farm. I started them out real little. I used to rotate the boys on the tractors. I felt that the monotony of sitting there, rocking, would put them to sleep, so I would let one sit for two hours, then take him off and put the other one on. They were only about eight when they started that. They've always been real responsible. We felt that we had to treat them like adults so they would act that way. I think that was an important part of their growing up, giving them responsibility at an early age. The girls didn't do the same kinds of chores as the boys quite so early, but they can drive tractors, work on the huller, or run a machine.

Because I was raised in the Depression, I didn't have anything to speak of, and I thought, "Well, I want my kids to be something." I always taught them, "If you are going to do something, do it well." If they were going to school, they should be good students. Well, they just fell right into a natural kind of schedule — they didn't complain. They had 4-H one night a week. In high school they were on the basketball, volleyball, softball, track teams — almost everything. School plays, class officers, student body officers — they tried everything.

I always worked very actively, and I took them with me when I did things, and pretty soon they started helping me. If you've worked together as parent and child, you learn to trust each other. When Dan was a junior in high school he bought a pickup with his own money earned through 4-H and working on the ranch. Every day he took the girls to school. He didn't have to, but they all just enjoyed getting ready for school and going together. I think farm families tend to be close-knit, because you are constantly working and creating something together. We do so many things together as a unit on the farm that our interests are very much in common.

Dan and Mike are obviously brothers, cut from the same freckled, sandy-haired mold. Mike is a little stockier, Dan a little taller, but both are ready for action in faded jeans, muddy workboots, sweatshirts, and red farm caps. With only two years between them, they are so much on the same wave length that they nearly finish each other's sentences. Their honest, open faces and thoughtful responses to questions make them the kind of young men that one would like one's daughter to marry — no fears here of deception or negligence.

## DAN AND MIKE BREIDENBACH, AGES 25 AND 27

**Dan:** I was born in 1953. Ever since I was old enough to hang on to a tractor, I wanted to be a farmer. We've both been riding around with Dad ever since we could remember. Except for school, the farm is where we spent all our time. We're quite a ways from town, and there's not too much else to do. I never thought about doing anything else. I considered taking a job after college and working for some company away from the farm here for a few years. But I never got too serious about it — seemed like it was a convenient time to come back to the farm and start in, so I did, right after school.

**Mike:** I was born almost exactly two years earlier than Dan. Unlike Dan, though, I never really made up my mind on what I wanted to do. After I finished college I worked in southern California for two years for a large corporate farm that was owned by an oil company. I lived there two years, helping to manage their farming operation. It was a good experience, maybe, but I like our life here much better. It's more free and easy-going here. The high pressure down there got to me. You had to produce to warrant your being there, and if you didn't, you were down the road, so to speak. It never rained there, so you never had any time off, and being single, I didn't get any opportunity to meet anybody except at work, and I worked 30 miles from where I lived. It made for a real long day.

**Dan:** Dad never encouraged us to stay on the farm. It was up to us what we wanted to do. He said, "Go to college, stay as long as you want, and when you get done you are welcome to come back if you want to, and we'll see what it looks like then."

**Mike:** Up until the last five or six years, we worked in the summer and if we wanted to, certain times of the year. But a few years ago the farm expanded quite a bit, and real quickly. Opportunities came at certain times. If you're able to do it, you can't pass up opportunities, or at least you shouldn't if you plan on staying in the business. My father and uncle had some good years, and they bought some land, and they started more intensive farming. Now it's more crops and more work — but it's also more money. When we were starting high school it was mainly just dryland farming. It wasn't really a possibility that everyone could come back to the farm. But when we were down at college, they expanded, and then they needed more help. By that time, too, my dad and his two brothers didn't want to be out there all day every day — they still are, that hasn't changed a bit — but at least they wanted to know that they didn't have to be. So then there was room for us to come back, there was opportunity.

**Dan:** For a long time we were the only employees working, really, along with our cousin. The uncles kind of managed it, though the three of them worked too. No money ever left the family, really. Our cousin is farming with his father now. A year and a half ago the partnership was split up three ways, and we are one part of the original Breidenbach brothers' operation.

After high school I went to college at Davis and graduated with a Bachelor's degree in agricultural economics. I got a Master's in that field too. I figured I could learn to grow sugar beets or wheat or nuts up here, but I wanted to get a different perspective on farming as a business. I took some finance classes and chose to study the economic side of agriculture. The theory classes, themselves, are not something you can directly apply. I haven't sat here after I got home from college, plotting all my marginal cost curves, and then raising that much production of sugar beets or whatever. But it complements real well what I do around here. I can't pull it straight out of the book to help me decide something, but it's in the back of my mind when I make decisions. It helps me farm more intelligently. I took a marketing class, an appraisal class, several management and finance classes. Every one of them was a winner.

Mike: University work helped open my eyes more to the whole picture from one end to the other. My world is a little broader than just the farm. Down at Davis, everybody I knew was from the city. It was a completely different world from here, where everyone is the same. I was (I hate to say) somewhat narrow-minded before I went down there. I am glad I went to a large school to meet people of different backgrounds.

When I worked down south for the corporation, farming was totally different. Everybody there, large or small, has a longer day. Farming takes more time, though you have better results. The land is better, yet you have to watch things closer because costs are so much higher. They emphasized the bookkeeping end of it much more — it was all computers.

Down there are a lot of really big farms, and some smaller ones. Public versus private corporations, really. A public corporation is one where the stockholders, the owners, can be anywhere in the United States. Then there are the family-owned corporations, like us. In a public corporation farm, you are responsible to someone you never see. The only time you see the owners is when they are not happy with you. After a while you get to thinking about the bad side all the time. As long as you do your job okay, you never hear anything. But if you don't do something right, then you hear a lot, and it is always negative.

These big firms aren't as efficient. There are really smart people in there, but one guy cannot handle as much as they want him to. And sometimes people are not quite the caliber they should be all the way down. While I was there, the firm really expanded and took in many new people. This really created a turmoil, because everyone wanted to know where they stood in relation to everyone else. After a while it started bothering me, because everybody else was so worried about it. This sort of thing mushroomed. Finally they had so many problems they just decided to sell out.

Employees of these corporations are not farmers, they are managers. The firm might not even own land. Sometimes they farm land for somebody else, maybe with a lease-purchase option. They're really speculators. After buying, they might keep the land for a maximum of five years and try to sell it for more than they bought it for. Hopefully, they break even on farming. But the farming end of it is not the chief reason for getting into agriculture. Land investment is a tax write-off for the oil company.

**Dan:** Talk about efficiency — they are not like we are. They have too many people working for them. They really don't give a damn. They could care less if they get something done, or if they do a good job, or if something is working right. If something breaks, they just keep going. Employees have no personal interest, they are just on the payroll.

A few years ago some of these large corporations got into agriculture, but they are getting out now, because there is not the return on investment in farming that they can get somewhere else. If they are a public corporation, their stockholders are saying, "I want to make 10 percent on my money." But agriculture is characteristically less than 5 percent — maybe 2 or 3 percent — return on your investment. My dad could make more money if he sold this place and invested the money somewhere else. He would have, on the average, a higher annual income.

**Mike:** Corporations have advantages. They have lots of capital. They can have nice equipment — tractors lined up from here to there. But there are too many people who are not actually contributing anything. If you are large enough that you need 15 bookkeepers, why, then, the people out actually doing the farming have to pay the bookkeepers' way too. You get too much deadwood in the organization.

I have been on both ends of it. We were all spread out, from Bakersfield to Fresno. When I got there they gave me a map about a foot-and-a-half long with all the roads marked on it and said, "These parcels are where we farm, go out and find the places." I would spend a half day on the road. I had a two-way radio. We were so big we had to have two channels to reach everybody. We talked to people every day that we never saw for six months, except maybe at a Christmas party.

**Dan:** Right now we have only one field that doesn't touch the rest of them. Advantage? No time on the road. On my pickup I put only about 4,000 miles a year. It's not the main pickup, of course. Dad's pickup and Mike's are newer; if we need a part, they run to town to get it. But in my pickup I drive only about a half-mile down here to the tractor, and then at night come back. But when Mike worked down south, he had to get up an hour early so he could be out to the job on time.

If my dad were to work 40 hours a week, he'd get done on Wednesday. That is the living truth. Even at this time of the year, January, we may work tonight or tomorrow night in our orchard. There is never a dead month because we are diversified. We have done that on purpose.

**Mike:** There used to be slack times, but we changed. Now we work like crazy around Thanksgiving, Christmas, to get all of the grain planted. Sometimes now we double-crop. There is pressure to produce up here, but it's a different kind of pressure. Down there we had pressure, but we had to put it on somebody else. I was not allowed to drive any tractors. If they saw me, they told me to hire somebody to do it. Yet I know I drive better than anybody that we could hire. Being smaller-scale here, we know we can get done. We can work at nights, we can overwork for a couple of days, and than take it easy for a day or so. Down there you couldn't recuperate at all. Somebody had to be there all the time. Even on Sunday I went out there.

**Dan:** The pressure we feel to keep busy now is the pressure we put on ourselves. We ask for it. We don't have to drive tractors, but we choose to. We grew up that way. My dad and my grandfather can hardly hear — they got deaf driving tractors. We are about half that way too. But doing the physical work is a way of life. We can do it better than anybody else. Down in Bakersfield we didn't know what we were going to grow until just before we started planting. They had to look at the <u>Wall Street Journal</u> to see what the futures looked good on. They changed it up to the last minute. But we make choices like that on the family operation pretty much too. We changed crops this year — more grain and less row crops — because in the ones that we grow, prices have been down. But we make the decisions, nobody makes them for us.

**Mike:** In our operation Dad is still definitely the head, but we discuss decisions more than we used to. We generally do what he says, because he has more experience than both of us combined. He is the boss. Once in a while we come up with something, and he will change his ideas a little. But generally what he says still goes, especially since he owns most of the land!

**Dan:** Dad has made a will. I don't know what's in it, I haven't asked. But we all know the family land may be a problem with the 160-acre limitation. That is our biggest worry. You can't make a living on a small acreage around here. It is a pretty tight squeeze between cost and profit. You have to be on top of things, and have to be a pretty large size to make it work, in this county.

As far as large corporations coming up here, there is not much land for sale. The people who own aren't planning on getting rid of it. A large corporation would probably be looking for several thousand acres. Hardly anything that size comes up for sale. There was one corporation place — overseas money — down here at the county line that just went under. Together five big farmers took over the ground so the company could pay its debts. But generally there is very little turnover in land up here. Down south, they are still developing land, it is still coming into production. Up here it is all more or less developed, all pretty well established. The owners aren't in the real estate business, they are in the farming business.

**Mike:** Our biggest concern is the 160-acre limitation on the federal canal water coming here next year. We are in a bad groundwater area. The wells we use have large motors and are real deep — 900 to 1,000 feet, and the water table is around 300 feet. It costs a penny to get a gallon of water up on top. So we really need the canal water. But if the 160-acre thing goes through, the ownership will have to be broken up (not necessarily the farming units, but the ownership). If they enforce that law, we will have to buy the land from our dad. Maybe our two sisters might also. And it would have to be at current prices. No shifty deals; our dad can't give it to us.

**Dan:** We hire Mexican workers off and on. Contrary to a lot of white people who want a job for a week or so, so they can take their paycheck and get drunk, the Mexican people come up here to work. They are the best workers that ever were; they will be there every day. A lot of them go back to Mexico during the winter months. Plenty of illegal aliens work around here too. That is just the way it is. The Mexicans, legal or illegal, come up here to work, and they really don't care what the job is. They will do anything. You can't find white people that will move sprinkler pipe. We end up doing it ourselves if there aren't any Mexicans around to hire.

**Mike:** My sister and I are nine years apart. When I was in school it was 10 percent Mexican; it is 30 percent in her class. If you look at the younger grades, it is about half. The Mexican workers are settling here with their families now. The entire state is like that. The Spanish-speaking are not really mainstream yet. They really stick together, but things are changing. They are doing better all the time. They all have such large families, though — if they had smaller families, they would be better off, financially speaking.

**Dan:** If you don't speak Spanish on the farm, you can't manage the workers. We both speak Spanish pretty well, learned it in high school and college. I knew I was going to need it, and I do, every day. But I think the workers play dumb to a certain extent. They understand some English. If we didn't know any Spanish we could still communicate. Dad mostly waves his arms and points.

I am kind of a laissez-faire person. The most efficient farmers are going to survive. That's what is best for the consumers if they want low food prices. We talk about the middleman driving up costs, but it is the consumers who want their dinner wrapped up in a little foil package that they can pop into the oven. Maybe the middleman by advertising has made them desire that kind of packaging with their food. The middleman performs many more services than he did before. But the consumers demand it now.

Mike: The way things are going, there will be more and more control over agriculture. The idea of breaking farms up into 160 acres is to make it possible for more people to farm. But people don't realize what is involved in farming. They talk about the homestead days 100 years ago — I heard that from people at college. But if they try it, they usually don't want to do it for more than two years!

Dan: A man and wife and a couple kids can subsist on a small farm, but they are not going to contribute one thing to consumers. And there are a lot of people out there who need to eat — too many people. Homesteaders can subsist on their own, but they will not have anything to sell of any quantity.

Mike: But we have been out of college for a while. I don't know if thinking has changed. In the 1970s things were pretty wild — Vietnam and everything. People seemed to want to have their own piece of land, and Cesar Chavez would come by Davis all the time. Everybody thought he was great. I would go listen to him. He would never say anything! Even if I hated the man, I could never say anything against him, because he would never say anything.

Dan: He has done a lot for farm workers. But like the Sierra Club, we don't always agree. But if all these folks weren't there, why, we might not pay enough attention to some things. Everybody serves a purpose. But it is getting to the point where farmers have to make a little noise. This morning Dad was up at a meeting in Chico. Some people want to outlaw all agricultural burning in the valley, and they wanted some farmers to show up. I think we are going to have to get more involved in farm issues, and more involved in organizations.

How do I feel about farming? It's too much work! Still, I don't have to show up tomorrow, if I don't want to. But I know I will.

Braceros (contract workers from Mexico) harvested sugar beets in the 1940s with hand tools. Reprinted with permission of the Department of Special Collections, University of California, Davis, Shields Library.

# ALTERNATIVES TO THE STATUS QUO
○
## The Weidner Family

At the end of a long lane threading through plowed fields is a complex of tree-shaded buildings and animal corrals. A brick sidewalk leads from the asphalt parking area to a modern ranch home. From the big front window of the tastefully furnished, comfortable living room is a splendid view of the "smallest mountain range in the world," lying just across the Sacramento River.

John Weidner, wearing cowboy boots and gentleman's western-cut shirt and pants, sits at an antique poker table with pistol shelves beneath the top. The atmosphere is gracious, prosperous. John is well-built, trim, articulate, and sure of himself. Occasionally his slight drawl breaks into almost boyish humor. He likes talking, is full of detail. His wife Vivien supplies volumes of coffee and a lunch worthy of a fine health-food establishment: homemade triticale bread, fertilized eggs, a seafood salad plate. She is small, attractive, warm, and very committed to John.

JOHN WEIDNER, AGE 56

I was born in 1922 in San Francisco, simply because my mother's doctor was there. My grandfather Weidner originated in Schleswig-Holstein. In the 1860s the Prussian armies took over and conscripted the young people. In 1867, when he was 17 years old, my grandfather hopped the boat to evade that draft, came to San Francisco, and up to Colusa County because he knew some German people here. He went to work on a farm and gathered enough money to buy 160 acres of land, which he cleared. The first year he had a good crop. He worked diligently, and two or three years later his employer, Mr. Schultz, rented his ranch to my grandfather along with mules and the seed. He hit another big crop that year, and that got him started up the ladder. He mortgaged his land to buy some more land a little farther west. And that's how he did it. He was a grain and livestock farmer.

In 1902 my grandfather bought three small farms together in this place where we now live. He built the house that I was raised in, a two-story house with five bedrooms and thirteen outside doors! The story is that he had eight children, and when he said "Scat," he meant "Scat!" So there were lots of doors to get out of, and they are all eight-foot doors, with thirteen-foot ceilings in the house. Two families live in it now, families who work for us here on the ranch.

I was an only boy with two sisters, one older and one younger, and went to the local schools. They had just dissolved the country schools around the area and integrated them into Hardin. We had a good school system, good teachers, had bus service right to the door. I was in 4-H and raised chickens from the time I was nine years old. That was my enterprise for a long time. As I got older I was supposed to milk the cows twice a day, before school and after. When I'd bring the milk in — we didn't have any electricity in those days — the cook would put the milk in pans and skim the cream off. And then it would sit in there and curdle and I'd have to haul it back out and slop the hogs with it. That always made me kind of disgusted. I'd get up a little late and I wouldn't have time to change my shoes before I'd go to school, and then the manure would be on my boots and the kids would tease me. So I didn't like milking cows.

I've had a horse ever since I was three years old. I've always ridden and loved horses. I'd rather stay home and drive a team in those days than go to school. In fact, I can remember I stayed home and mowed hay with a team rather than go to Tahoe on our senior sneak in high school.

My mother was a strong disciplinarian. She used to break a stick over me once in a while. When my father did raise up in his wrath, I knew it was time to kowtow, but he didn't do it very often. My father was a gentle man. He went to Cal Berkeley where he was a roommate of Earl Warren [who later became chief justice of the Supreme Court] and a classmate of Bob Sproul [who later became president of the university], about 1909-10. He was going to be a lawyer. But then, just like my sons, he came back to farm with his brothers. He liked football, and he continued to socialize with his fraternity; he liked that life. And he liked poetry and liked to paint. He was very social. He didn't really get into farming heavy-duty. He liked it, but as a way of life rather than to make money. My uncle Charlie, his partner, lived in town. He came out and did the planting and the harvesting, but otherwise we didn't see much of him.

We had a Chinese cook. We fed all of the men at the ranch who worked here in those days. My mother didn't cook, but she kept the house and the yard. She loved to garden. She sang in the choir and at funerals. She was a very entertaining and a very energetic woman who made jokes out of everything. We had a real happy life.

In high school I took agriculture and shop and the basic college requirements. I was in Future Farmers of America and in a few positions of school leadership. I played all the sports. School was easy, I didn't study hard. I went to Cal Poly after high school, majored in animal husbandry and minored in crops. When I graduated in 1942 I joined the Air Force. I became a bomber pilot, then an instructor, and eventually an instrument check pilot. Everybody who came from overseas back to the United States had to go through my squadron to get checked out. I never went overseas. After the war, when they didn't have the money to keep people flying around, we got discharged.

Vivien and I came back to the ranch. My folks had rented the land out for rice during the war because they weren't farming themselves. I thought that as long as they were renting it, they might as well rent it to me. So after a debate (they didn't know if I had the ability to farm, and would be able to give them the income they needed for the two partnership families — plus our family, Vivien and me), I had to prove myself the first year with 50 acres of beets on very uneven land. It couldn't be row irrigated, it had to be contoured. So we farmed beets as if they were rice, with checks in the fields. I brought in the first beet harvester in the area to try it, a mechanical harvester that wasn't too successful. We also raised 300 or 400 acres of grain — I think that year it was barley. And I worked on the side, custom work, for different people and made some income.

The next year I went into rice, a couple hundred acres, and maybe 130 acres of beets. Everything just grew. After two, three, four years, I had kind of weeded off the other tenants, taking over more and more of the family land. My parents had the equipment and I rented it from them, that helped too.

We grew rice on the ranch here from 1939 to probably 1953. For two reasons we went out of rice. We're on an alluvial plain here, with sandier soils suitable for several crops. And, because of the livestock fences, rice on one piece of ground would force alkali up into another piece of ground, and so we were continually ruining some fields to generate dollars on another field. We quit rice and took out all the fences, sold off all the livestock, and I leveled everything in the early 1950s. I then went into heavier row-crop farming and different rotations. For 20 years we had no livestock. When I first took over farming this ranch, there were oak trees scattered over it, there were washed-out gullies from the 1916 flood, and alkali holes that you had to pull through with chains in the wintertime because you couldn't run an engine through them. Since that time we've leveled and releveled and releveled; and put in irrigation systems and concrete ditches and wells and pumps. I have been regenerating the land and diversifying the agriculture.

Rice grows well on very heavy soils, but if you have an option on what to grow, you might as well grow something else. It happens, though, that rice actually saved this ranch during World War II. During the years of the Depression, and the dryland farming that my father and uncle did, the ranch got so run down there just wasn't any income coming off it. There were the wet years in the mid-1930s, and then the dry years when there was no price for grain. But rice came in 1939 and from then through 1943 it brought the money back in.

Of course, during those years the prices were fixed. We had the Office of Price Administration and the Steegle Amendment. The Steegle Amendment provided us with farm pricing in World War II, at parity — that's where parity comes from. In all those years of the 1940s, we were at a minimum of 90 percent of parity and a maximum of 110 percent. Because of that, we generated enough wealth from agriculture in those years to pay for the war itself. Look at the presidential economic reports and you will see our debt structure in those years was very minimal compared to what it is today. Those were pretty good years for farmers.

We have loam soils here mostly, since all this land by the river is on the flood plain. On the Storie maps these are very good soils. We do have some high sulfur content, and we always watch out for alkali. In the dry years we had, you could see the alkali taking over more because those deposits weren't being leached out. But this is highly productive land, if it's cared for. When I took things over, the ranch wasn't in all that great shape, but I've always tried to keep up with new thinking about agriculture, and over the years I've made many changes. I suppose I'm typical of other farmers around here as far as production is concerned, but not typical in my thinking. I'm probably credited with an A- or a B+ for being innovative.

I inherited the home ranch here in 1963, after both my parents had died. It had passed down from my grandfather to my dad and then to me. The farm now is operated under our name, Vivien and John Weidner, but we do have a corporation, Weidner Ranches. So part of the holdings are family corporation and part are husband and wife proprietorship. We both have wills, and we leave things on our deaths to our three boys.

Currently, the tax dodge is to skip a generation with a life estate for the wife. So some of the land will be left that way and some will be left to Vivien clear, so she can make the decision if she wants to sell it, rent it, or do whatever she wants.

Before we took title to the ranch, though, things had been going downhill for farmers all through the 1950s and early 1960s. Eventually I got involved in farm politics. Starting in 1952 during the Eisenhower administration, farm policy shifted to what they called "sliding scale parity" in the Benson era. Sliding scale meant that each year we went down to 90 percent of 90 percent parity, until our farm prices reached a bottom in the early 1960s. We were losing farmers all the time. By 1968, when we were being offered $1.95 for wheat at the ranch and it cost $2.25 to produce it, I could readily see that I couldn't stay in business, and no one else could. That's when the National Farm Organization (NFO) hit California, in 1968, and that's when I got involved.

In 1955 NFO had grown out of all of the protest movements and the "holiday" movements in Indiana and Iowa. NFO then was just a bargaining organization. They didn't even block production — they were just going to hold for a price. That didn't work, so they had "holding actions" where they wouldn't sell anything, and they would get their neighbors not to sell anything. It created a lot of animosity in the Midwest and it wasn't very successful. I didn't know that when I got involved. In 1968 we joined because they told us they would block our production and get contracts. They told us they were doing it back there.

Jack Hunter, a neighbor, first told me about it. He got wind of it from a fellow who came out to talk to the Peach Growers Association in Stockton. There was a meeting in Hardin at the Firehouse. Jack Hunter believed in NFO. He was a young farmer at that time, I think only 35. He needed assistance, everybody did. He tried to get me to go to the meetings but I wouldn't go. I didn't think I needed it. I had been on the Land Bank board for 15 years and I knew how financing was. We knew what we were doing. We could survive as well as anybody.

But when $1.85 wheat was the market, I realized something had to be done. When NFO had a meeting to organize a chapter in this county, I went and listened, and then joined. They elected commodity committees. I was on the grain committee, and they selected me for the chairman. I figured if we were going to be involved, we had better get doing something. So we blocked a lot of grain and put the price on it that we had to have — $2.35 — at the ranch.

This is how it worked. Everybody who had grain — let's say I had 1,000 tons of milo, and my neighbor had 500 tons, and another guy had 200 tons, and maybe another guy only had a truckload — would sign in together and agree to sell all together. That's called "blocking." We created this block of grain, about 15,000 tons, and put the price on it. Then I started negotiating with the grain brokers I had always done business with — and could not get them to pay for it, naturally. This was in September, October; we had to wait finally until January or February before we got our price, but we finally got it. As the market moved up and we were the only ones left with any grain, they had to have it. But the funny thing I learned is that as soon as we sold that block of grain, the price moved up again — but we didn't have anything left to sell. So then and there I started realizing that we had to be in the market all the time and sell part of our block at a time and move the market ourselves.

In the meantime, nothing was getting any further organized in California. Vivien and I were talked into going back to Iowa to see what was going on with NFO. When we got there, we could see that they weren't very well organized — Oren Lee Staley and the administration in Corning, Iowa. Originally they had sent people out to California to get people into their organization. But after we joined nothing happened. It was primarily a self-help program. Vivien and I stayed for three days. She organized their files in the grain office. They talked me into taking a salary on a part-time basis. I said I would, to get it going. So we started organizing the state of California then — the regions and the marketing committees. We had a state marketing committee and started blocking production, making contracts.

I did a tremendous amount of traveling and talking, day after day, night after night, all through the state, in 1969. I was received well, because we were already successful farmers. We weren't going broke ourselves. I was able to convince larger farmers to do it, such as Tyrone Black in Shannon, probably one of the largest grain growers in the state. We became very good friends, and he became the first president of the NFO.

Finally, by 1971, I had offices in Raleigh, in Hanford, in Williams, in Clear Lake, and was coordinating offices in Washington, Idaho, Utah, and Arizona. In California we probably had 5,000 members. I was at this time a national staff member and the first director of NFO from California. Out of the structure we organized, we blocked production, ending up with about a third of the grain in California. Finally, through trial and error, we convinced Continental and Cargill, the two major buyers of grain in the world, to do business with us on a contractual basis. This was reflected back into the Midwest, and those people got with it, and the program got going. We were very successful — in fact, we had so much money on the books that that was really the downfall. The administration did not have the ability to cope with it on a business-like basis. We were moving a quarter of a million dollars a day through the office. The problem was not getting professional business management soon enough.

Then the Russian grain deals came in July of 1972. We were holding for a price on some grain contracts in California and couldn't get it. Then, all of a sudden, the third or fourth of July, Continental's vice-president called me up and said, "We want to buy that contract." I couldn't get it ratified because everybody was gone for the weekend. We made the sale, after talking it over with certain people, on the basis that it was what we were asking for. Well, on the fourteenth of July the word came out about the Russian grain sales, and the market moved upward — and it moved so fast that it moved way past our contracts. Came harvest time, our growers didn't want to deliver. We were stuck with the negotiated price, but we couldn't deliver. When you are short on delivering a contract, you have to pay the dollar difference. Then the buyer goes to buy the grain somewhere else. That hurt. The national movement of NFO broke right there.

It was a historical accident. All of a sudden, all of these millions of tons of grain went to Russia, raising the price so much that the previous work on the contracts was all to no avail. It was too bad. There were ways of getting out of that position, but Oren Lee would not listen to my ideas. I knew if we took the grain these farmers had, but wouldn't sell to fill the contracts, and got them to agree to block it into a higher price level, we could make those sales. I knew it would go even higher, because we were in that period. If they had done that not only could we have sold the grain and made the farmers money over and above that original price, but also we could have paid off the grain companies for the excess dollars over the contract. We could have gotten out of all that within the period of two months in October, November of that year. But Staley wouldn't do it; he had a "crisis" drive, getting members to volunteer to pay the money up to pay off the grain companies. He finally did it — he has a wonderful way of convincing people to part with their money. But it broke up the organization. At this time in Colusa County you couldn't get anybody to participate, they were so disillusioned with that turn of events.

NFO is a cooperative under the Capper-Volstead Act. But the difference between this cooperative and others is that when you join a cooperative, at least in California, you turn over production. Then that organization markets it and pays you a percentage at harvest, and then retains 6 to 10 percent over a six-year period to operate on. You get payments from them —

but you have no say in what they sell that product for. Whereas in collective bargaining as in the NFO, the theory is that you negotiate price by generating a block of production to contract — and you then bring it back to be ratified by a majority vote of those producers, so they have a say in that contract's price. It's more actively making a price than just taking it.

The idea of collective bargaining is moving the market — not just selling the product, but moving the market to generate the cost of production plus a profit price. You move the market by being on the market all of the time. When we would sell a block of production out here, then the nonmember down the road would want what we got — or a nickel more (because a nickel more buys a farmer). So he would be more demanding of his brokers. That moves the market. It works, I guarantee you. You give me 20 percent of the production in the United States, and I'll get you any price you want for it — including the rocks out here in the street. You block enough rocks together and I'll get you a price for them; I know how to do it. There are other people like me who know how to do it too.

I still belong to NFO, but I don't participate. NFO is dormant. Lots of people are just waiting for something to happen. NFO is trying to hold together the structure, but they don't represent enough production to be effective. During the interim time since 1972, though, they have built up their computerized bookkeeping system and know enough to hire qualified people. But I don't think that the farmers in the Midwest, but especially in California, are willing to follow the national leadership back in Iowa, because they don't have faith in it. Still, there are groups in the Midwest blocking production together in large volumes to try to make direct sales overseas, while directly bypassing, hopefully, the major grain companies.

My experience with other farm organizations is varied. I have been a Farm Bureau member ever since I started farming. We get our insurance from them. I don't personally agree with the political stands the Farm Bureau makes. I found out early on that the Farm Bureau, along with the Cattlemen's Association, was part and parcel of getting rid of the Steegle Amendment. The history of the Farm Bureau has been to organize farmers, but not so they could price their products.

I'm not happy with the national Cattlemen's Association either. The new board has got three members who are related to the King Ranch of Texas. They are all part of the Deltex Corporation, which is tied directly up with the Rockefeller families in Argentina and Australia. They are producing beef to be imported to the United States. So we have a hard time with the dilemma of how the import laws are written. The Cattlemen's Association and some of these other groups are really dominated by agribusiness, no question about it.

We went into the cattle business again in 1973 for two reasons. I'd just seen the NFO dissolve for all intents and purposes into an ineffective structure, and I was really upset about that, and needed something to be more occupied with. Number two, realizing at that point in time that my dream of solving the economy's problems through organized agriculture wasn't going to work, I thought we'd better take care of ourselves. So we decided to become self-sufficient on this ranch.

We got our chickens, our horses, our cattle back on the ranch. At that time they were bringing in what they called exotic cattle breeds, new breeds, to the United States. I took a hard look and found they had a lot to add to our Herefords, Angus, shorthorns. As I studied it more, I decided that what we needed was hybrid vigor. The same thing we'd been doing in our seeds had to be done with our livestock. So we went into buying exotic cattle. I took a tour to Germany, to see what they were doing over there. I was very impressed with their breeding programs; they're very well-organized, and their record keeping is excellent. So we're in the Gelbvieh cattle business now, and in a crossbreeding program through artificial insemination. These Gelbvieh cows are the old work animals. In Europe today they still use them. Sometimes you'll see them, one or two cows hooked up to a wagon, going out to a field. They milk one or two or three cows, but they don't keep a bull, because the cows are all artificially inseminated by the government in the herd program. They're gentle and have strong legs because they're raised on concrete. So if we ever have to worry about energy, I'll have the cows I can hook things up to if I have to.

Then I went to the feedlot system so I could recycle the manure. Smells like the devil, but we haul it off every so often. In time I have in mind to have a 500-head lot on each ranch, where the manure would cycle into a pond and pump into the irrigation system and go back out onto the soil.

We have a hill ranch and we're doing some experimenting with planting grasses on the range. We're going to put a full-sprinkler-set system on the whole 160 acres. The canal's just come in now, they just finished the last reach. When they pump that up there, then we'll have irrigation water. It'll cost me about $600 an acre. People ask how can you do it with cattle, but I can show here where we're running 500 head of cattle on 135 acres of irrigated pasture, and by-product feed, behind the wheat, behind the beets, behind the corn, and I generate that many dollars from those cows on 135 acres.

Besides the cattle, we've got horses. We went to Skylark Farms, the number-one pleasure horse breeders in the United States, for mares, and bred back to their studs. They show our horses — we have the number-two Green Novice mare on the West Coast this year. The ribbons are in my office. We're going to make some money on horses, but mostly it's fun and something I always wanted to do.

Actually I'm phasing into retirement. The boys are farming. They took over the first of January. Right now they're farming 3,000 acres. We own half of that, the rest is rented from family and neighbors. The crops are 660 acres of sugar beets, about 400 acres of corn, and about 1,000 acres of wheat. Alfalfa is 500 acres, seed crops about 400 (melons and cucumbers). This year we have 123 acres of triticale, which is contracted to go to Egypt. Right now we have about 500 head of cattle, but we're selling down. We've got 262 head up for sale.

Every day I'm out there checking the land, the livestock, the labor. The boys call me the resident advisor. I try not to interfere with their decision making, but there are times when you have to guide them.

I learned a long time ago that I can make much more with a pencil than I can with my back. I used to run engines 12, 14, 18 hours a day when we first started farming. Then I got over that and started hiring labor. When you have labor, you have to supervise it, but more than that, you have to plan what generates money for you and for them. A lot of my time is spent managing things so our employees have better conditions to work with, so they produce more. I think you have to cast your shadow on the ground on every crop every day, seeing what is growing out there, seeing the conditions of the soil. For a long time I always carried a tape recorder in the pickup. As I drove around I could take notes. Now we've installed radios in all of the pickups for communication. We are spread out over 25 miles between the farms we run, and the radios save time. We always try every day to get more efficient.

I've seen farmers going out of business — educated young farmers who didn't make it and went bankrupt. They wanted to get bigger and bigger, thinking if they produced enough volume, they could make up for low prices. Their economic education within the university has been Keynesian economics, where they were taught you could borrow money and spend borrowed money and not have to make a profit — in other words, substitute debt for profit. Well, it doesn't work. It's like taking an airplane out full of fuel, flying out over the ocean, and somewhere out there you get to a point of no return, where you are closer to where you are going, then you are to where you came from. What happened to those boys is that they invested so heavily in equipment for the tomato business that when the price on tomatoes dropped, they faced bankruptcy. I urged the ones I am thinking of not to do this, but to live within their budget. But they said, "No, we want to spend the money for this equipment and we are going to get big, and we want to do it while we are young and energetic." As it turned out, it just broke them. That was in 1969 or 1970. If they could have just hung on for two years and hit the Russian grain thing when the prices moved, they could have made it. But in the meantime they were broke, out of business.

We've survived through good times and bad for a combination of reasons. Number one, I farm with a pencil. If it won't work on paper, it won't work. I figure everything out ahead of time. I estimate my income on the low side and my expenses on the high side. If it works on paper, I'll do it. Number two, we have always had farm credit. We have always been financed that way. I knew if I lived within a budget I couldn't go far wrong. I've been very careful with record-keeping and cost analysis. Even though we take risks, we don't take them too big. Then, too, over the years we've had enough cushion that we have been able to absorb some losses. We were fortunate to have the land. I had the chance to buy my sisters out at the original land cost, rather than an inflated cost. And we have always made the equipment stretch as far as we could. We still own a 1950 Chevrolet truck that I use and am proud of. I'm going to keep it running as long as I can. We have built much of our own equipment.

We've kept up with changing technology. The major technological change in sugar beets was mechanical harvesting and thinning. When I first started farming, we harvested the beets with Mexican labor — or any kind we could get. We had Navajo Indians one year. But mechanical harvesting changed the whole picture. It allowed us to winter-over beets and have a spring harvest. With beets overwintering, we get two or three tons more sugar to the acre. A beet doesn't make sugar until it's used up all the nitrogen in the soil, and then the production goes into sugar. Otherwise, the nitrogen goes into starch. And no question about it, it's much cheaper to harvest mechanically. With mechanical thinning and the type of seed — pelleted seed — that we use today, we can almost plant to a stand now, compared with all the thinning we used to have to do. We still bring in a hoeing crew, but instead of having 100 people working for us now, we probably have about 20.

Of course we've got a lot heavier equipment now too. The more horsepower we've got, the more we've been able to go down and break up subsoils, ripping two feet deep, and we landplane with an 80-foot plane. It makes the land smoother so it irrigates better.

We're finding now that soil compaction has partly been caused by chemical fertilizer. We found out over the years with using anhydrous ammonia and the different chemical fertilizers that our soil particles were getting less adaptive to irrigation.

Water wouldn't penetrate, and the soil particles were sticking together in heavy clods that wouldn't break down into seed beds. We had to have bigger and stronger equipment in order to break those clods up. We had gotten very far away from the old livestock type of fertilizer and laying the ground idle for a year or two in the different crops or pasture. We were working the ground too much, too hard, double-cropping. We wanted to get this soil into production where we weren't dependent upon the corporate chemical industry to survive, and we realized that we had to go back and utilize manures. We add minerals to this manure and compost it so it will make an organic fertilizer that will go onto the land and produce nitrogen. For mineral additives, we use lots of calcium, magnesium, and other things that make this compost conducive to creating bacterial action in the soil. It's windrowed out, and turned, and added to. This is the first year that we've done it here.

We feel that we're stewards of the land. We don't own it, but we have to leave it better than it was when we got it. All the wealth in the world has to originate from the soil. When we get into that concept of economics, we realize that we have to protect the soil, to keep it for ensuing generations so they will have wealth too. In the last five years we've been going more and more toward an organic kind of farming to build the soil back up again.

We've experimented with "Biochem," an algicide that was produced in Arizona originally. You fly it on, and it opens up the soil and gets the bacterial action working. And that lets the water penetrate, breaks up the soil particles. Four years ago we put Biochem on half of each field on several different crops and harvested them separately and went to the bank with them separately. And on sugar beets that year we generated $235 an acre more where we added Biochem. That convinced me. The problem with Biochem is that the people producing it weren't businesslike. They finally ended up going bankrupt. When we wanted to talk about Biochem with the university they wouldn't even talk to us, because they're getting money from the chemical corporations and this isn't a chemical corporation deal. I found out about Biochem when I was director of the NFO. I was traveling over the United States talking to different people. I became friends with Dr. Loren Busch of the University of California, who raises three-quarters of his food organically in his backyard in Berkeley, with daisies in between for the bugs and the whole bit. And there are people who have done this for years.

Eventually the costs of these procedures for commercial farms will be lower. Right now it's about the same or a little higher than conventional methods. Actually we're losing our chemical fertilizer plants in this state. There's only one producing in

California today. We invested originally in Valley Nitrogen, which is a farmer-owner co-op. It grew so big we put a plant in at El Centro, and bought a gas deal up at Red Bluff, and bought the Hercules Powder plant, but now all that's being dissipated. They're selling off. They've stopped producing, because anhydrous ammonia is coming in from Mexico and Russia. Chemical fertilizer is being brought in from outside the United States, just as your shoes are, and your textiles, and all your electronics — because it's cheaper to produce elsewhere.

But it's going to take a long time to educate people to the value of organic fertilizer, I know that. I don't care, they can look at me as a nut if they want to, but when I go to the bank, I know what I go with. You can't convince people with pictures of growing plants that look lush. You've got to prove to them with what you take to the bank — they'll pay attention to that. The profitable education is really the only one you retain.

As far as pesticides are concerned, we told Ralph Nader when he was out here making a documentary that we have to use them. Because if we don't, there won't be any food on your table. And that's a sad thing. But the thing is to be darned careful, use them correctly, and don't use them any more than you have to. We had a terrible grasshopper infestation here last year, and we had to use several kinds of pesticides on them even though we didn't want to. We lost one stand of beets. So I can't say that we can raise food organically and not use pesticides at this time. But I think there'll be a time when we will.

Rotation of crops is one way to contain pests. One thing on this farm that we've done wrong was to concentrate our crops — beets in one section over here, corn over there, wheat and so forth, and rotate that way, because it was conducive to irrigation, conducive to movement of equipment. But by the same token, when we got a disease in one field, every field adjacent got the same thing. The grasshoppers or the lygus spread. If we have these fields separated, with interim crops in between, we have a better chance of stopping that. So we are working on breaking the fields down into smaller sizes. We've numbered every field and keep the records on each field. We do a lot of record-keeping. We have to be careful with our sprays not to kill our bees, ladybugs, our good bugs. We're trying to figure out how we can do a better job without using pesticides. We'd rather have the ladybugs eat the lygus, or we'd rather destroy the eggs before anything ever comes around to become a pest. With our 2-4-D and with DDT we've created monsters. As the mosquitoes created new generations, they became less susceptible to being killed by that chemical. We'll have to go biologically at the problem, but it's going to take time. And this is where organic farming comes in. The insect bites and lives on the weak plant. A strong plant can sustain itself, despite that insect. When you get your land balanced, it produces a strong plant.

Not only is that strong plant less susceptible to disease and pests, it also produces stronger food for you to eat. We grow our own wheat, and I try to keep the wheat for us to eat from the fields that didn't have as much chemical fertilizer. Vivien grinds our wheat and she makes our bread. And we have found out that we don't eat as much food, we don't have to watch our weight as much as we did. That homemade bread seems to give us just the energy and drive that we need. Our vegetable garden is organic and has been for ten years. We just use manure on it and don't put on any sprays or chemicals. We've got it enclosed with wire to keep the chickens and peacocks out. We just feel we eat better.

In the NFO we got into a triticale program, which didn't work because it was pushed onto the market too soon. The varieties hadn't been perfected enough, and the markets weren't there. I talked to the people at General Mills, and they said if they had enough quantity of quality triticale, then they would make a flour out of it and sell it. It's good for you, no question about it, and it'll feed the world in time — it's a high-protein grain.

There was a lot of experimentation with organic production in 1973-74. Out of that has come Chuck Walters, Jr., an interesting man who is the editor of Acres, a writer, and an economist. He's put lots of writers in his magazine like Dr. Ulbricht, who's the father of organic farming, and C. J. Fenszau. Some of them are kooks and some of them aren't. You have to weed them out.

What are the reasons why people are so unresponsive to these things I talk about? Well, the Committee for Economic Development (CED) has been around since the early 1940s. I went through the Library of Congress files on it and it originated with the same group that started the Committee on Foreign Relations (CFR). On the board of directors of CED are 200 people. They're all presidents and vice-presidents of the leading corporations of America. The only farmer on there is Bob Kleberg of the King Ranch in Texas. Yet they have these economics programs developed for all the facets of the economy, including agriculture. I have copies in my office somewhere. The problem with agriculture, they said in 1962, was that there was too much waste of resources, mainly human. And the answer to this waste of human resources was to divert them into urban areas and make them city workers, to move those people off the land. They had to do this several ways. One was by changing the definition in the census report of a farmer. The other was to mechanize. When the cotton harvest was mechanized, the negroes were all moved into Watts and elsewhere, to create a social problem.

CED meets periodically and has a staff in Washington, D.C. They function behind the scenes all the time as advisors to presidents, whichever political party is in power, it doesn't make any difference. Standard Oil has somebody, and Bank of America, and Chase Manhattan. I'm not saying "conspiracy," but I think that down through the years they have exercised a hell of a lot of power. That's a fact.

I consider myself a constitutional patriot, which is kind of a wordy thing to say, and I believe in what our founding fathers tried to build here. I believe in the Constitution and the Declaration of Independence, as it was written, but not as it's been interpreted by different people. I do believe we have to have a feeling for our fellow human beings and our neighbors. We have to work together. Any time that one group thinks it can gather up all the economic power for its own betterment, then we're in trouble. Politics gets into why we can't get organic farming across, and why we're importing cheap fertilizer.

There are trends in California agriculture that I am concerned about. We have already lost, for all intents and purposes, our electronics industry. We don't produce any more TVs in the United States. We have lost our textile industry almost completely, and half of our steel industry. Now Bethlehem Steel is going to build a $2 billion complex over there in China. We have lost 65 meat-packing plants in California in the last two years. In the United States today, we are losing one meat-packing plant per day. We are concentrating the processing of beef in Iowa, Nebraska, Missouri, and Texas into the hands of four or five meat-packing companies who are going to box the meat and ship it all over the country, plus importing whatever else they can. If I were in the butchers' union today, I would really be up in the air, because they are going to lose their jobs. They are using processing structures and machinery back in Omaha that our obsolete California plants can't compete with. So Allen Pack in San Francisco has shut down and is buying Iowa meat and filling its contracts in California stores with Iowa beef. Labor is part of it, transportation is part of it. It is the whole structure of the economy. Resources are concentrating in the hands of fewer and fewer entrepreneurs. Right now you can't get a hog killed between here and Los Angeles. You have to go 500 miles to get hogs butchered.

There is one easy solution to all this: Have the courage to have complete economic control. The problem of today is debt. We have been substituting debt for profit. Inflation is nothing more than paying today's bills with tomorrow's money. In order to reverse that, we have to go back and get things in balance. Stop spending, stop borrowing, and create — balance our production with our costs. To do that now, with such a disrupted economy, I can't see any other way than to go back and index everything as we did during World War II under the Steegle Amendment. There should be a simple law, just like the minimum wage law which says I can't hire you for less than so many dollars an hour: You can't buy any products from me for less than 100 percent of parity at the farm gate. This would create new wealth, earning power, and generate the dollars that substitute debt for profit. It is the reverse of the present system. We put it on the Department of Agriculture computer in January of 1978. It showed, even without all of the proper inputs in there, that in two years we would start to stabilize the economy and slow down inflation. In five years the economy would be balanced again.

Now consider Colusa County. We have 400 or so farmers in the county. This is an agricultural county, so the farmers are the ones basically concentrating the wealth. At the same time, we are almost 50 percent minority — very few negroes, mostly Mexicans. The state by 1982 will be over 50 percent Spanish surnames. You can see it in our schools. This county has never changed population — 12,000 in 1900 and 12,000 now. We like it that way. But out of the 12,000 we've got 1,400 now on welfare, and 2,300 on unemployment, and we're bringing in almost 50 percent Mexicans. We need to bring in agriculture-related industry, to create jobs in the economy.

There aren't many opportunities for new farmers. The land sales are going mostly to larger farmers who are expanding or to foreign investment. The young boys are not buying that land, not at $3,500 an acre, because they can't ever pay for it in their lifetime within the current price structure, unless they get some kind of outside money. Our boys are lucky, because they have a base to start with. There's no question that they love the land and they want to farm, but it's a question of economics. And a question of whether their wives can settle into the harness or not. At least 50 percent of farming is a good wife, if she backs her husband in what he has to do.

VIVIEN WEIDNER, AGE 56

I was born in 1922 and raised on an almond ranch one mile south of town. My grandfather brought the first almonds into northern California. My father mainly had almonds, although we always had several thousand chickens and a small acreage of grain. I had two sisters, one older, one younger. It was very much a "work ethic" family. We gathered the eggs from the time we were very young. My mother had to supervise the messhouse for the men, because in almond ranching we had many, many people in the summertime. We had apricots too when I was very young. So there was cutting of the apricots to do, and working on the almond huller.

My mother managed the ranch herself after my father died in 1944, for about ten years. Then she rented it out because of her health. Farming is not exactly women's work. But the renters weren't as good managers as she had been herself, so she sold the ranch in about 1970.

I met John in first grade, but we didn't go together at all until after high school. I graduated in 1940 and went to Berkeley for two years in prenursing. Then the war started. Just before John finished training we decided to get married. We thought we would be together for about six weeks. All of the cadets then were being sent overseas immediately, so I planned to go back to San Francisco Medical Center — I was all measured for my uniform. But John was kept in the training command in the United States, so I stayed with him and never finished nurses' training.

In 1945 we came back here. We weren't necessarily expecting to come back to his family's place. It was really open; he also considered staying in the airline service as a pilot. But we grew very anxious to come back, if possible. John's father and uncle were farming together, so it wasn't a clean-cut operation where John could just step in behind his father. He didn't go into the partnership, though. He was always absolutely on his own. He rented the land and the equipment from them. So even at a very young age, 23, he was doing his own thing.

Over the years, I've been very involved in the daily operations of the ranch — not like a midwestern wife who might be getting on the tractors, but most supportive. I have helped with the books, paid the men, paid the bills. Early on, we didn't have a bookkeeper, though now we do. I have been the premier errand-runner, of course. I go out to Colusa or Oakfield or wherever to get equipment, supplies, or parts. And I have handled the correspondence, and I've helped evaluate which companies to purchase from.

I haven't been involved with physical things on the ranch. If we had chickens or something like that, I would take care of them, but that's incidental. I have always had a garden. I think it is very important to have fresh fruits and vegetables and foolish for a country person not to have them. We've always got squash, tomatoes, broccoli, cauliflower, chard, strawberries, eggplant, onions, peppers, and rhubarb — just whatever. We have a home orchard too. We don't get much off it because John puts his steers in there and they keep attacking the trees no matter how we try to protect them. But we do get the walnuts and apples. I love to make jellies and jams. Sometimes I can fruits; and I freeze corn and squash. So garden things are a year-round part of our diet. We make our bread too, for daily consumption. That has all started within the last few years. We grind the wheat from the ranch to make the bread.

There has been a change with the family's feeling about diet. I never used to bake bread, though I could have, and we needed the money then more than we do now.

A farm wife has to be very supportive. Farming is such a hassle — it always has been and is becoming more so. If a wife thinks of herself only with her own immediate needs, it is almost impossible for a man to farm. She has to be sympathetic, because the farming absolutely has to come first, even before the family. There is no way a farm can succeed today otherwise. Husband and wife have to be a working team.

When we started farming we didn't have much money — we had nothing but hope. But we just knew that we were going to succeed, and it never occurred to us that we wouldn't be farming the rest of our lives. Today when young people start farming there is no such assurance. They worry from year to year. Are they going to get refinanced? There have been great changes. It is much more difficult for young couples starting out now, especially if they start out without a broad base. Young people think differently now too. When we were first married, it didn't occur to a woman not to consider her husband's role as absolutely primary. Today those concepts have entirely changed. Maybe rightfully so, but I think that it will compound the difficulties. There are not just economic problems for the family farm today. Roles within the families have changed too. I sense that with the younger generation. The team aspect is perhaps not so readily accepted. Farming is hard on a wife unless she has had a farm background — or else is a very wise girl. A city girl is used to the 9-to-5 operation with the weekends off, and the three- or four-week vacation her father had. There is no way a farm can operate that way.

We never encouraged our sons to stay in farming, because we went through some very difficult years. We felt if they had an alternative way of life, it would be beneficial. Then they could come back to farming if they cared to. In any event, they all three have come back for various reasons. Tony, the oldest boy, always planned to farm; he took agricultural courses in college. Our second boy went into landscape architecture. After he got his degree he got very good jobs. He was a Robert Trent Jones golf course architect — he loved that. But he didn't like living in the city and in the corporate structure. Our youngest boy graduated with a B.A. in Liberal Arts and he too has come back. For different reasons they have all come back. There has been an attraction here, and it hasn't all been economic.

As we have started to make a transition in the family here, my husband was very wise: He decided that if the boys were going to learn to farm, they would need his guidance, but they needed to be making the decisions. He allowed each one to rent about 500 acres. Each would go in a different direction: The oldest boy would grow sugar beets, the middle one would grow seeds and beans, and the youngest one alfalfa, so they would not be in each other's way. After several years we found out that that would not work, because they were pooling the equipment and pooling the men and pooling the water. It was hard to decide who would get which water, and which men, and when. Just in the interest of survival (even though we didn't like anything about partnerships), they went into a very structured partnership, working with a lawyer. Even while they were making preparations for making it work, they were making arrangements for dissolving it if it wouldn't work. There are very clear steps on how they can get out of it, if it becomes intolerable. Actually, the boys to get along amazingly, even though they are very different. The problems will come only if their wives make things complicated.

There seems to be a gap between farmers 50 and 60 years old and those 20 and 30 years old, at least in our county. I think that occurred for economic reasons. It wasn't that the boys now roughly in their 40s didn't want to farm, but the opportunities weren't there. We had a kind of farm recession at the time they would have been approaching agriculture in the 1950s. They went off to the cities and to other occupations. Many of the boys now returning in their 20s have all kinds of degrees too, and the opportunities and the brain power and the will to do other things, but some preferred to stay in farming, and opportunities in the early 1970s looked promising. So they came back — and some are hanging on by their fingernails now.

We came in the 1940s. It didn't occur to us that we wouldn't succeed. My husband has a very positive nature, he only understands success. There was some stress, of course, for farmers during those years in the 1950s and 1960s. But we were very diversified. When John's father and mother passed away, and we had the opportunity to buy out his sisters' shares, he doubled the rented acreage that he was farming in order to make those cash payments. We just didn't spend on other things, we were very frugal.

I don't know of any farmers who really feel rich. There is always another tractor to get, or something. City folk think the farmer is rich because he has so much land. But you can't eat land and you can't spend it — yet you have to have it to produce income.

John has always farmed in a very business-like fashion. He always had a reason for everything. He didn't just do it because his father did it. He was maybe ahead of his time in that way. The boys do have a somewhat different philosophy about how the land is treated. It didn't occur to us to farm in an organic fashion. John always felt he should leave the land in better condition than he found it, but he wasn't concerned about chemical fertilizers. We didn't realize they would do the things to the soil that they have. The boys are learning that there are better ways to farm. And they have influenced John, too. We are very receptive to the things the boys can offer. But if he feels the boys aren't doing some chores on time, he will be very vocal.

It's a comfortable stage of life for us right now. We know it isn't going to last forever. At 56, our good years are limited for doing the things we would really like to do. But we have made three trips to Europe. My father died at 52. I realized early that a man was nice to have around. That is one reason John and I have had this good rapport. You have to live while you have the chance, because if you wait until everything is perfect, it will never happen. We will probably travel more. John's chief interest is the cattle now, while the boys are taking care of all of the farming.

Even though we have this beautiful life, I worry about the future. I am really frightened for the little country town. When I was in grammar school there were four or five grocery stores in Hardin — and one by one they have gone out of business. We have many empty stores on Main Street. My cousin owns some of that property and has a hard time renting it, even to someone who might want to put in a craft shop. We have lost our drug store. We are forced to go 15 or 20 miles even to get a spool of thread, because the little store in town has no selection.

Everything now seems to revolve around the social life in the Firehouse. In a little town the rural volunteer fire department is a very important cog. I hate to see that possibly coming under the control of the state. I like to see local control.

About 15 years ago they unified the schools. I think that was a very bad mistake. Our community fell in line, and all of our little schools have closed. I don't think it has improved anything, and it has created more problems. Some of those little schools produced some very bright and outstanding people. In our little community here we had a very good school. All of the young people graduating, seniors who went on to college, did exceedingly well. When you come from a small community, you are used to having some leadership roles — being the officers in your clubs. Everybody has a chance to be a person, they know who they are. They are not just little numbers. I just hate to see the bigness growing everywhere — we lose so very much. There is less community life than there used to be. Television has had its effect — people stay in to themselves more.

I worry about the future of the family farm. I have read reports predicting that by 1990 agriculture is going to be totally incorporatized. What is going to happen in these next five or six years? What kind of disasters are they going to create?  The only reason anybody sells a farm is because he is economically forced to sell.  Nobody leaves the farm because he feels there is a better life — there is no better life.  We are really frightened.  Family farms represent certain values.  If they disappear, it's going to be a real loss to society.  And the price of food is going to skyrocket.  The corporations have always been able to price their products, and they will keep on reaching for control.  We will be getting peanut butter that says "Made in Japan."

> Buck and Laura are newlyweds.  Buck is thin, intense, full of portents and politics.  His words tumble out almost feverishly at moments.  For such a favored young man it almost seems strange that he should feel so strongly that the world is conspiratorially stacked against him.  But he is passionately committed, in love with ideas, fervently pursuing his goal of self-sufficiency through the uncertain haze of ominous world events.  Laura is quieter, her blue eyes not so worried, her voice soft and sometimes hesitant as she gropes for the right words.  Farm life is ahead of her.  She knows she will have to make adjustments, but she is happy to be part of this family, with all its hard work, long hours, and heady talk.

## BUCK WEIDNER, AGE 27

I was born in 1951, and I grew up here.  When I was one year old, my parents built this house.  The values that I grew up with are, I think, old-fashioned.  I'm not religious, but I believe in the basic Christian ethic of what is right and wrong.  You don't lie to anybody, you don't expect them to lie to you.  If you promise to do something, you do it — that basic ethic.  I don't see it any more.  Maybe that ethic is more common with farm families, because they are more isolated.  The temptations aren't there.

We lived nicely in the 1950s.  Parity was 95 percent, and the farmer could make a profit.  But when I was in the middle of high school, in the early 1960s, the sliding scale of parity was finally starting to take effect and the farm recession was starting.  You don't notice the early part of a recession in a farm community.  You borrow your money and you pay it back.  But you are not paying back as much every year.  During my last years in high school, the NFO came along and then we really started finding out what was going on with the economy.  My dad was really involved, and I was ready to tag along.  I'd go to most of the meetings, and I became very political.

I didn't always assume I'd be a farmer. I was a good student, and in college I went into Liberal Arts. There were minor clashes between father and son. When you grow up in a farm area, you know you can't start a farm unless you have a silver spoon — or at least bronze, or something. To start out, you've got to have some kind of help. If you choose not to go along with your parents, you might as well choose another profession. My attitude toward coming back to the county was, "I won't be back here unless I am on the farm. I won't work in a store, like lots of other people who have failed. I'll come up with another completely different career so I won't have agriculture look me in the face and say I couldn't make it."

I went to Europe in my junior year and bought a lot of paintings and was going to be an art dealer. But I didn't really want to live in the city. It was in Europe that I made the final decision to come back to the farm. I took the train up through Germany, through the Rhine Valley, and saw all of the small farms there, and I thought, maybe I don't need as big stakes as I thought, to come back and be a small farmer. So I came back — and a "small farmer" turns out to be 3,000 acres! Right after graduating from college I returned and started working on the ranch.

We had sort of an agreement that it would be like a postgraduate course — I'd work for Dad for about two years, training for farming on my own. It was tough. Dad's last experience with teaching was in the Air Force. When you teach your troops to fly a bomber, it's life or death. It's life or death out here too, but not in the same sense — it's profit or loss. Some of our clashes occurred again. Luckily I was 22 instead of 14, so I was able to take it a little better. The first year or so was pretty tough coming back, because I had been away from home for four years. But basically our economics were the same.

We're part of a group of friends we call the "Slough Club." This goes back to the NFO days. It's a group of farmers in this area, our neighbors, who were all in NFO. For some reason, these three men living right next to each other were the strongest men in the state. Our neighbors, the Shaughnessys to the north and the Hunters to the south, were involved with Dad. They had a kind of magic show that they would take up and down the state — Jack introduced them; he had the philosophy. Ed Shaughnessy had that Irish accent that nobody could resist. And Dad knew how to do it. Dad was the first director of NFO in the state. He sold all the grain for the western United States.

Everybody else before that had said, "What will you give me?" to the grain buyers, and consequently they lost money. Now back in 1914 and even in 1948, farmers were getting $3.00 for a bushel of wheat. In 1968 they were offered $1.83. Yet everybody else was getting more for their products. The rest of the economy had gone up 300 percent and farmers had gone back 50 percent. That means the rest of the economy was 600 percent beyond us!

We all believe in the raw material formula. All new wealth comes out of the ground. If you put $1,000 on a poker table, you will just trade that around as long as no one brings in any new money. But a farmer is going to put new wealth on the table. There is no other wealth besides what comes out of the ground or out of the sea. The farms, the fisheries, the ores, and the lumbers — those are the raw materials in the formula. But lumber takes a long time to reproduce itself, and ores are what is in the ground, from whenever the earth was formed. Fish are still pretty much what we gather in from the sea. But meat and vegetables and food from the farm are produced every year. And for every dollar produced in agriculture, we get seven dollars of new wealth in the economy. Right now farmers are only producing 60 or 70 percent of the wealth they could. The economy is losing that other 30 percent, and so we are looking at inflation and depression, the exact state of affairs in this country since Eisenhower decided to give us a sliding scale back in 1952. Take a loaf of bread: 3 cents goes to the farmer, but 70 cents is what you pay in the store. If 6 cents went to the farmer, it would still only be 73 cents in the store. The extra 3 cents doesn't amount to much, but it would keep the farmer at a high percent of parity, and keep the raw materials flowing in.

In 1972 the NFO got top-heavy back in the Midwest, and Oren Lee Staley tried to finalize his dictatorial powers over the organization. The Californians led a move to oust him. Dad was nominated for president. About ten people were kicked out because of their threat to Staley. Dad was still a viable candidate, but he didn't win. When that happened, everybody lost the chance for having the organization ever become a functional one. There were a lot of negative things going on with Staley in power that were never resolved. As it is, Staley finally had to resign seven years later, and the NFO organization is nowhere now. It was once a rising star to take over the whole agricultural scene, but it became nothing.

I'm very interested in politics and economics, but why get shot? I am completely against what the Rockefellers believe in. Not just the Rockefellers but the Mellons, Hunts, and the others on that Tri-Lateral Commission and the CFR. Read <u>Captains and Kings</u>. If you don't go along with the game, you might as well not be in politics.

I'm scared because I have every reason to believe that there are going to be only 250,000 to 500,000 farmers by 1985. I feel the revolution might hit. But if it happens, I am going to be a revolutionary. Scott and I were talking about it the other day. The problem is that we can't really define the enemy. The oil companies are a logical enemy: But is the person who works for an oil company down on the level that we see my enemy? The salesmen, my neighbors, and the friends I went to school with, are they my enemy? My premise is this, actually: The enemy is ourselves, our own lack of morals. We let ourselves get into debt because we are a credit-card society, not paying as we go along. I don't want to borrow money to farm. But here I am — I don't own anything except pieces of equipment that we have been able to purchase as we went along. I just barely got refinanced this last year. And we made a profit! If you make a profit and you can't get refinanced, it's rough.

I'm not sure where our problems come from, if they come from Washington, D.C. or New York City. There's a conspiracy, all right. Every time the Tri-Lateral Commission and the CFR have a secret meeting, they are deciding what is going to happen with the world. And CED is part and parcel of the same thing. Power is exercised by a relatively very few significant powerful rich men. They don't have to do much. They can just push a button in Zurich or New York, whatever city it is that they're having their meeting in.

We have an attorney-general trying to push a law to ban agricultural burning in California. What better way to knock off the rice industry? That's the biggest industry in Colusa County. For 100 or maybe 200 farmers, rice is the sole thing they can grow on the land they have. It's terrible soil, it's a quagmire. The only other thing they can do with it is maybe raise some cattle. Well, that might wreck the balance of the beef industry. Our beef imports from Australia are already putting ranchers out of business. Maybe the agricultural burning issue is part of the environmental movement. But I think most of these social movements are sponsored by the CED. Those people are brilliant people. Get the ecological movement, get the women's movement, get this and that; capitalize on dissatisfaction. Dissatisfaction is how you manipulate people. The socialist movement played labor against the rest of the people.

In any business where you have a labor supply, you have to deal with all of those people's emotions. So business gets rid of them. In farming, mechanization makes life easier. Farming is a big enough hassle. Dealing with people makes it worse. On the other hand, every time you get rid of jobs, you put that many people on welfare. Many jobs on the farm require unskilled labor. If the farmer makes a profit, he can pay labor. But the farmer gets caught in the money bind, and often the only place he can deal with that is to cut off labor. The banks find it totally justified to buy a new piece of machinery, if you are cutting labor. When I was getting my new budget together, the bank said, "Cut out labor — it's too high." I just don't have the money to hire people as far as the Production Credit Association is concerned. As a young farmer, the pressures are on me to get rid of labor. I can't make the decision myself. Actually, it would be easier for me and my brothers to farm the ranch ourselves. I don't even care to hire anybody, but I know part of my role in society is to hire people, because I have a means to do so. It's a quandary that we have to live with.

We are a family. We ask advice from Dad, but my brothers and I make the final decisions. We are basically going more organic, getting rid of the anhydrous, which has been foisted on us by the chemical industries. We found that our ground has been tightening up, and we are not raising the same crops that we were able to raise before in terms of yield. We were putting on 100 units of nitrogen and anhydrous and were only getting 60 sacks of corn instead of 80. Dad has always kept records of his crops, so we can document that yields have been going down. We know we have a hardpan that we didn't have before Dad started farming. That is what anhydrous does. We have had to get heavier and heavier equipment to work the ground deeper and deeper, just to try to break up that hardpan. But now we are going to compost from cattle manure and humic acids and humates, and we mean to get away from chemical fertilizer. The organic program we are following is laid out by a man in the Midwest named C. J. Fenszau, who has documented proof of its success. We will still make a profit with going organic. We will have just as good crops, if not better, by improving the soil, by making the soil in the plant root zone more permeable, by making the bacteria and the elements in the soil more readily available to the roots. When you put anhydrous in, the roots are not able to get their molybdenum, manganese, and stuff like that. Those trace elements are one tiny part of all the life that's in the soil, and just as important to the plant as the 100 units of nitrogen. If you don't have the molybdenum, even with the 100 units of nitrogen, you don't get a crop.

If I could really create the ideal situation for myself in farming, I'd like to have a couple hundred head of cattle, and raise crops to supply them with food — alfalfa, corn, wheat. I'd want to be pretty much self-sufficient. Barter for what I want. I'd like to be left alone.

I know sons in my generation who have not been able to come back to farming. I've been lucky. My best friend is in the army now because his father didn't buy the land that he farmed. People buying land now are the Arabs and the Italians, the French and the Japanese. In the European countries that are going socialistic and communistic, the landed people are getting their money out of the country and putting it here in

California and in the Midwest. In this state foreigners have a Californian buy the land for them. My neighbor's brother-in-law is a major buyer for foreign people. He buys land all over the country.

There is a good piece of ground up for sale not far from here, but we can't pay $2,200 per acre and make a profit. You have to look at it as if you were paying rent. You can't really afford to pay over $100-$150 an acre for crop rent if you pay cash. Wheat doesn't even create that kind of profit — it's in the $50-$100 range. If you get a 20-year note on a piece of ground at today's prices, you have to pay $300 an acre for it, note money. That is just exorbitant.

Dad ought to be around for a while, so we intend to rent for the next few years. We will inherit a certain part of his land, as long as we are not complete fools and make him mad enough to hate us. But inheritance tax takes a huge bite. The whole tax system is a farce, it just allows the government to keep growing. Inheritance tax ought to be banned. The government has no right to control the land, or the water, either.

I got married a few months ago. I don't think farmers could function without wives. I don't really enjoy butting my head against the world, when there is a conspiracy to put farmers out of business. It's hard work to go out every day if you are not doing it with somebody or for somebody. I was going to quit, because I didn't see any reason to kill myself farming, even though it is what I want to do. I was going to go down to Tahiti or some island, and read, write books, be another Gauguin. The farm is an ulcer factory. The elements are against you, and the government, the people in the community are against you, resenting the fact that you have land, or are richer than them, supposedly — even though they make more money than you and they don't even know it.

Out there is one of the prettiest sights in the world. I've always loved that view — it was my life's blood. But it still wasn't enough. Laura and I found each other just at the right time for me to stay here.

I went to Washington, D.C. last year, knocking on the doors of all of the congressmen, and talking to them about farm concerns. There were 30,000 of us back there. Scott and I opened the office in this county for the American Agricultural Movement. I was the first publicity chairman, calling the news people to tell them there were going to be tons of people in Sacramento for the tractor-cade. But you get pretty tired of trying to be a leader of nobody, and people resent you.

My father and I share political views, and my middle brother too. My oldest brother just lets life go through him, but that is okay. I think I would be willing to take up a gun and fight, if I could find the enemy. And I would be willing to burn all the crops if the banker decided to take the farm. I'd pour poison on every crop around here. I wouldn't let people take away what is mine. I know there are really going to be some grim times ahead. There is going to be a depression, and people will starve. Maybe people will want to come out to the farm and take the food. If they do, I'll shoot them. Or I'd be glad to barter — you pay me a turkey and I'll give you part of a beef. If times get tough, there will be a true exchange system. There will be no more debt.

You watch the price of gold; see the manipulations. We are trying to play the gold market, we think at least a currency like gold will come out of a depression. Somebody will want it, when the times get tough. Krugerrands, too, the South African dollar. We're trying to plan ahead to defend ourselves against hard times. It would be foolish to think times will get tough, but just say, "Oh, let it hit me."

We three brothers are up to our necks in debt, but at least Dad is still liquid. Our goal is to keep him that way. That means the land will still be here. Even if we go bankrupt, no big deal, as long as the family can hang on to the land. We will be okay. But I don't know how we are going to do it.

## LAURA WEIDNER, AGE 30

I was born in 1948 in Wilsonville, about half an hour from here. My parents moved to Sumner and I grew up there, on 25 acres in the country. My father was manager of the mosquito abatement district, but he was also a bit of a farmer. We raised some sheep and cattle, and had vines and fruits. I worked some outside, but mostly inside helping my mother. I wasn't much agriculturally inclined.

I went to school in Davis and took a degree in Child Development, and then at San Francisco State took my Master's. I worked in migrant education child care in Merced, Stanislaus, and San Joaquin counties as a coordinator for the child care programs at the migrant camps in the three counties. I decided I wanted to get some business administration background to increase my productivity in education, so I went to the University of California at Riverside and did the Master's program there. Then I went to the multicultural project at Riverside Unified, and after a while moved to Oxnard in Ventura County, where I taught child development and worked with the community there to develop programs in child development. I work now as chief administrator of all the preschool programs in Riverside County.

Last summer I met Buck, and we got married in October. It was a whirlwind courtship. We are sort of in an interim period, since I am still working in Riverside during the week, but flying up for weekends at the ranch. I'll continue working until I get the project done down there, but then we are thinking about having a family. I expect, though, when I move up here, I'll find some kind of work. I may get a doctorate, or work at the university, or at one of the colleges. Maybe I'll see what this area needs in child care, because this county is not as advanced as Riverside. We have a house, and I'll be part of the community. During the next few months I'll be phasing out of what I'm doing now, making the transition to a new way of life.

Marrying into a farm family is different. Having just turned 30, I had decided that if I didn't find someone I really loved and cared about, that I wouldn't get married at all — let alone marry a farmer! I hadn't anticipated that. My family thinks it's ironic that I am now married to a farmer after my whole life experience. But the Weidners are not typical farmers. They are a sophisticated family, aware of what is going on in society, aware of political happenings, and they are farming in a much more knowledgeable manner than the average farmer.

The farm itself, the business, is strenuous. The time Buck and I will have together will be minimal, because they work really hard. The hours are long. Now we have weekends together, but from October 21 until a couple of days before Christmas, Buck worked seven days a week. We would go to dances on Saturday night, and he would be all tired out, and Sunday morning he would be back at work at seven. So that's how life is going to be. That is another reason that I would want to continue with what I am doing, professionally, because I don't want to be a nuisance to him either. They do work so hard. I feel I'd have to have an outside interest of some kind to compensate for the fact that he is so involved in what he is doing.

I have had a lot of experience with migrant education and migrant children. When I worked in Merced, Stanislaus, and San Joaquin counties, there was child care in the labor camps. People felt sorry for the "poor migrants," but my feeling was that the farmers were doing the best they could. I knew grape farmers who were having a hard time providing homes for migrants and paying the wages. And the migrant families that I worked with made a lot of money during those months that they worked. As I got to know the families, they would sometimes invite me to come and visit them outside of Guadalajara and near Lake Chapala, where they lived in Mexico for the other six months of the year. It was not the traditional picture of the "poor migrant farm worker" that most people have.

Now there *were* people who lived in cars, who really were the "poor migrant worker." But in the Coachella Valley, farm workers are different from the popular perception. As I go to state meetings, people are starting to say, "Maybe the migrant is not what we thought. Maybe the migrant is really a six- month worker, who works real hard for six months, so that he can live the rest of the year on what he makes then." That is not such a bad life, considering that I myself work 11 months out of the year and have one month off. Health services are provided for the migrants, child care is provided, and so on. In a meeting I attended the other day, different terms were being used for "migrants" — the "mobile farm worker" and the "transients." The ones who are mobile plan specifically to move from place to place. The others just do the best they can. The larger population seems to be the ones who are purposeful, not the desperate ones. They have a relatively stable life-style. They move, but it is planned.

Now that I am a farmer's wife, I can see that to survive in agriculture you have to mechanize, because you can't afford to pay the wages to labor to get the product out. More people keep coming in from Mexico, because the Mexican green-card residents who have been doing the farm work are becoming upwardly mobile. They don't want to do the hard work anymore, the 12-hour days that are needed. But as agriculture loses those people, more people keep coming. Mechanization changes the process of what is happening. There are fewer jobs left on the farm, but the workers keep coming over the border anyway. The press is pushing the farmer to give the workers more money. If he has to do that, the farmer is pushed toward more mechanization. Yet even with mechanization, you still have to have workers. So it keeps going round and round.

Those who used to be mobile farm workers are often now settled out in the towns. More Mexicans are going on welfare. That has always been something they didn't do before, because of pride. In our child care programs we could count on the blacks to be on welfare, but you would never find Mexicans, because they wouldn't take it. But now they want to stay more in one place. In Los Angeles the population is half or two-thirds Mexican. The farmer then is forced to take the green-card worker or the illegal worker from Mexico.

There is also more separation and divorce among Chicanos than there used to be. Even in 1972 and 1973 there was practically no divorce — always the mother and father were working together. That is not true anymore. The Mexican mothers and fathers work together, but then they stay in California and live in town during the year, and they are breaking up. It's a big problem.

Why do we put so much money into welfare? Child care is supposed to be an attempt to help somebody make it. But it means you have to continue providing that service. If they get to that fine line where they start to succeed and are not eligible for child care anymore, then they have to make a choice. It is maybe easier to take welfare than to "succeed" and make only a little more money. I would rather see us put money into low-paying, entry-level jobs, and put money into child care to assist them, than pay welfare subsidies. We spend our money on foreign goods — producing them in the United States is what we need. It would be better for our economy if instead of paying for cheap goods bought from other countries we would spend the money on American-produced goods, and pay Americans to work to produce those goods.

The farmers that I have met are really sincere. They look at farming not only economically, but also as something of social value. Some farmers who meet in this county as a co-op say they feel a responsibility for the future of our society. They look at chemical farming; they look at wind machines; they look at different energy sources. Buck gets notices all the time. I attended one meeting that had to do with triticale. Their concern for nutrition is widespread. I was impressed. They really got down to it. Apparently the grain that is produced now does not have all the food value that's needed — if they don't change the grain, we will end up with bread that will eventually increase the retardation factor in children and decrease the physical prowess of people eating it. These farmers are saying that if they don't look at the long-range effect of the product they are selling they are not being responsible. A corporation does not usually have those goals. The people in this group are looking at not just economics and better production, but at how they can produce a product that will have a long-term positive effect on the American public. When I went to that meeting, I thought, "Oh God, I am going to a farm meeting." But I came out feeling, "Boy, you guys talk about things that are really important."

Large grain threshing crews were necessary in the days before harvest mechanization and bulk handling. Reprinted with permission of the Department of Special Collections, University of California, Davis, Shields Library.

# SOME STAYED HOME
## The Savely Family

This is flat, open country, where the wheat and rice fields are very large, and farmsteads few and far between. The Savely house is surrounded by a small walnut orchard that breaks up the monotony of the flat land. On this day it is raining torrents and the ditches along the sides of the road are brim full. Inside the house a bustling, energetic Mrs. Savely carries a pot of strong coffee from kitchen to living room. A small table is piled with family mementoes and local history books.

Genevieve Savely is an enthusiastic history buff and an incessant talker, keenly interested in family and community affairs. Gray-haired, smiling, casually dressed in slacks and sweater, she expresses great pride in her children, occasional impatience with her husband. ("He's just not a good businessman.") Tom Savely is tall, quite slender, looks younger than his age. In corduroy pants, a plaid shirt, and cardigan sweater, he looks as if he has experienced hard work and weather, but he is soft-spoken and pleasant. At one point he stops his wife's chatter with a look. He is much more reticent than she.

### THOMAS SAVELY, AGE 68

The Savely family is all French. The maternal side of my family settled in the Louisiana Purchase directly from France, probably in the late 1700s, before it was sold to the United States. One of the family members became a fur trapper for the Hudson Bay Company and was a guide to the explorers who came out to California before it was a state. He brought his family out here to settle on one of the Spanish land grants. The paternal side came from Canada during the Gold Rush. I speculate that the reason my ancestors left France was because they were probably third or fourth sons. One, they did not inherit a farm, or two, become a priest; or three, did not go into the army. That's why they went into the New World.

My family started farming on our present place in 1910. My mother and father came down here after they were married in Shasta County and rented a ranch of about 400 acres. I remember my mother telling that the landowner stayed out there and boarded with them, and my mother didn't like that at all. I think they were there five years. Then they bought 160 acres adjoining the place they rented and gave up the lease on this first place. That 160 was what became the home place, about 13 miles west of Plainview. Eventually there were nine of us children, and we made our living on this 160 acres. After some years my folks added another 100 acres, which made it 260.

I was the oldest son. I was born in 1910 out on the family ranch, and went to a little one-room country schoolhouse behind the hill. There was one year when only the Savely kids were going to that school. We had a big family, and the neighbors were pretty sparse. I had five younger brothers and three older sisters. The first four kids were closer together than the rest. Then they began to get spread out a little bit, farther and farther apart.

During my childhood we milked about 15 cows, and of course we raised hogs, chickens, and so on. I remember that when I was six years old, my dad brought home a two-and-a-half-gallon milk bucket, brand new from the store. He gave it to me and said, "Here, you are going to milk this cow. She is nice and easy to milk. You are going to learn how." So I learned to milk when I was six years old, and I milked cows all the time I went to grammar school and high school. We separated the cream and sold it. It was a business. The hogs were a business. Everything was a business. That's the way you lived. That was your income. Of course, we always had so many young heifers or steers that we sold from year to year too. That came with running a small dairy or whatever you want to call what we had. We farmed a little, but what little land we had was dryland. We made hay. We milked those cows until I got out of high school. There was no future in milking 15 cows, or anything like that. It was just a way my folks had for making a living and raising us kids.

I wanted to be a farmer from day one, but I wasn't going to milk cows. I always had the feeling when I was a kid that I wanted to be a farmer. I didn't want to go to school, I didn't want to go to college, I wanted to be a farmer. Of course, my dad was the boss, but he had all these kids and we got different ideas as we were growing up. We had stanchions in the barn to put the cows in when we milked them. One day Dad came home and us kids had taken out these stanchions. We just hauled them out. We said, "Dad, we are not going to milk any more cows." It was a revolution! "What are you going to do to live?" he said. Well, we just phased out those cows. We decided to just let the calves take care of the milk instead of milking the cows ourselves. So we started a cattle operation. The cows we milked weren't dairy cows anyway, they were beef cows. We had Herefords and mixtures of everything, and then we started raising hogs too.

This was a case in which the kids, as they were growing up, didn't see any future in dairying and got the parents to give it up. And tried to get new ideas into their parents' heads and get away from the old methods — which were all right for them, they did fine — but it was time for change. Kids get different ideas. They think they can expand and do better. So that is what brings about the changes.

There were six boys in our family and every one of them stayed in farming. I had one brother who went to college for three years, then worked in Oakland for a while. But he came back to the farm. The rest of us never left at all. When we got out of high school, why, we just stayed right here. In the beginning we were working for our parents. The folks were still the head of the family, and we weren't paid any wages or anything like that. At the end of the year, if there was any money left over, we divided it. This was in the beginning of the 1930s, during the Depression. Times were pretty hard then. I remember my dad coming home from town saying, "The banks are closed." Farm prices were really down. But we never missed a meal, because we raised practically all our own produce, outside of flour and sugar and a few items like that. The rest of it we had — we had our own pork and our own everything. We were really self-sufficient, so we didn't suffer.

We had some neighbors go through bankruptcy and lose their ranches. I just found some old records that my mother kept when we were kids, starting back in 1924. I came up with this old book the other day, out at the home place — I knew it existed, but nobody knew where it was. She kept all the books all those years, while raising all those kids. Every day she always kept the weather report too. It is really interesting, I will tell you.

Our neighbors just below us had 1,000 acres that adjoined us. They had a bankruptcy sale and sold all of their equipment, and we bought part of that equipment. Right in that book is the price we paid for all the items. These neighbors were just a man and his wife, with no children. He had a brother and together they had inherited a big ranch from their father. But evidently it was easy come and easy go, as the rule goes sometimes. The fellow just finally lost that ranch. It was that way with a lot of big ranches. They weren't frugally and economically run. They had more to survive with, but they started at the top. That is where the trouble was. We started at the bottom, with no way to go but up.

There was a lot of labor in my family. We didn't have to hire much. In that book it shows the labor we hired, and the amount of money we paid. My dad paid a dollar and a half a day. We needed extra labor only when we were cutting or hauling hay, just a short period of time. We raised rice in the 1920s, and of course the book shows all the rice labor during the harvest.

In the 1930s I was in my 20s. My next oldest brother went to college, but I stayed home. My younger brothers were all in high school down to grammar school. We all lived at home, and nobody was married. After I got out of high school, why, then we started to expand. Any place we could rent a piece of ground, we rented it. We had no borrowing power or anything like that, so our dad was the one who went to the bank. The bank said, "Well, you are a good honest man, and you have a bunch of kids, you are probably a pretty good bet. We will loan you some money." It was that kind of thinking. My dad was typical in that era of a farmer.

I will tell you what those farmers would do in those days — those people had no hobbies. In the morning you would get up and if you had nothing to do, you would go to town and stay there all day. Just to get the mail and to bring home maybe nothing, but you stayed there all day. All the old farmers, especially, would do the same thing. They would go to town, they would sit on the ledge of the bank window, all lined up there, and talk. All day. In the wintertime, at least, it was real sociable. The heavy work periods during the year were just putting in the crop, preparing the land, and cutting hay and harvesting a little grain. That was the extent of it. All the rest of the daily chores the kids did. Work was as the seasons came around.

We grew rice in 1924, 1925. I was just 14, 15 years old. It was the very beginning of rice out on the west edge of the trough here. They were expanding the irrigation district in the 1920s and that's when we started. We would bind the rice with a binder, pulled with horses. We would thrash it with the old stationary thresher — you had to have a big crew of men. We didn't have the stationary equipment, we hired that done. But we had bought a couple of old binders and fixed them up, and we had the horses. After we bound the rice, a big group of Hindu Indians would come in and shock it. It would stand up and the air would dry the moisture out of the rice, maybe three, four days. The biggest problem was that you had to wait for the stationary, because it was thrashing here and there for the rest of the neighbors. It was a contract outfit that moved from place to place.

The Hindus came from India into this country in the 1920s. When they first came they couldn't own or rent land. But Hindus knew how to raise rice. That was their specialty, irrigating and so on. The Hindus got around the land law by going through a white man to hold their lease, and that's how they got a foothold into rice farming. They would live in groups, a whole clan of them. We had a place out here that we called the Hindu camp, where a lot of Hindus lived. I remember going over there with my dad. At night they would be sitting around in a big circle, smoking a pipe, and would pass it all around, one by one. Just men, in that particular spot. The women came later.

The Hindus were the main source for rice workers in those days. But a number of white men would follow the crops. They would work in the rice, then go on and follow the grain crops in California, Oregon — and go right around in circles. We had men come back completing the circle each year. But it was really a loose system. It was word of mouth if you needed someone. You would try to find out through other farmers where there were workers. You would go to town and these laborers would come in, one by one, and you would pick them up off the street. They would be standing on the corner looking for a job. If they looked like pretty good workers, why, you would ask them if they wanted to work for you. You took a chance. Finally enough of the Hindus came in that they just kind of spread all over. The land law changed and now some of these Hindus are big operators. They have a temple over in Marion City.

In those days we planted the rice dry. We would broadcast the seed, like other grains, on dry land. Then we flooded it. That became a problem, because the black birds would come in and pick up a lot of that seed right off the ground. We had to start trying to sow it in the water. Then we would go through the water, pulling the broadcast, crawling over the checks. Oh man, what a problem! That changed right after World War II. As soon as the planes came in, why, everybody started sowing by airplane. It just wiped out the other method of sowing, because it was so far superior. Then you could plant it in the water, and that solved the bird problem.

A lot of this area was new land where you didn't have a big weed problem. Eventually, as you kept farming the same ground over and over, you would have more of a weed problem. That's one reason we used to plant every other year, and let a piece of ground lay off for a year. That would keep the weeds pretty well cleaned up. You could go in on the fallow ground during the springtime and work it up to kill all the weeds that had started. But now you can't afford to do that — you have to plant it every year.

Rice paid well sometimes, but it had its bad years too, I will tell you. A lot of this ground out here is marginal land. Rice is about the only thing you can grow and make anything off of it. They are very heavy soils and poorly drained, and mostly alkali. Rice can take that better than anything else. But those two years in the 1920s that we grew rice, we really didn't make any money. It just cost too much to hire all the labor.

Cultivation practices have changed. Of course this area's all row crops now and there's a different cultivation altogether. When we started out, we used to plow the rice ground. Well, you don't plow anymore, only once in a few years. You chisel instead, and disc it. We used to plow all the time for the dry farming, but we haven't used a plow now for years. Nowadays the discs are heavier, bigger, they dig deeper. In the early days they were small and light; they didn't do the job. The newer equipment has changed the method some.

They're learning more about irrigated rice all the time, especially over the last few years, than we knew years and years ago. I'll try to explain just a little bit. If you can keep the water shallow after you flood your field and sow your rice, maybe three or four inches deep, that's the ideal depth. We used to have to hold it six, seven inches. Well, nowadays our land is leveled, so we don't have high spots, low spots. We're using less water. That's better for the plants. The plant has to come through this water all the way, and if it's shallow, it gets through quicker. If it's deep, it takes longer for the little plant, the blade, to get to the top, and a lot of those blades, if the weather's too cold, they'll die, and then the plant has to start another one. In the early days, of course, they had no choice but to use more water, because the land wasn't level, there were high spots and low spots, and you had to hold the water deep to get it to cover, otherwise in some places you'd have the ground sticking out.

Leveling is our biggest asset. When I first started farming, the ground was just like the Indians left it. Leveling started just a little around here and there. You'd go out there and try to take the knolls down a little. There were commercial operators, but if you had any kind of equipment yourself, you'd work at it. You'd survey it yourself. My kids helped me lots of times.

When we first started raising rice, fertilizers, pesticides, and herbicides were unheard of. We didn't use them at all. We didn't know about them, and besides the land was all virgin land. We used to raise a 35-sack crop, on an average. Well, now you can't make a living on 35 sacks. We used to raise rice every other year, and then let our land lay off for one year. But since World War II you can hardly afford to do that because you have to pay the taxes on that land, so you have to produce something on it. So now every year you put it to rice. And then, of course, you have to use more fertilizer. You end up raising rice on fertilizer. The soil just holds the roots of the plant.

We farmed rice and grains, but eventually we got into more livestock too. We had to keep trying to expand. You can't just stop and back off and sit down. The grass grows under your feet if you do. Especially with that many in the family. If you have any gumption at all, you are going to try different things. We always had cattle. We sold calves at weaning time — we never fed anything out for meat. We first started with Herefords but we wound up with Angus. Then we went into the sheep business beginning the early 30s. We got started because we were able to rent a sheep ranch next to us, owned by an old maid and her brother. He died, so we rented this ranch from her and took over the sheep. Then we just expanded on it. I think she had 400 ewes. We built it up to about 2,000. That went on until 1972. We raised lambs for spring market. We didn't go in for any scientific range management. We just hoped there'd be rain and the grass would grow. As for brush control — we hoped somebody would drop a match sometime and accidentally burn it off!

There were a lot of livestock in the county at that time. Everybody was trying to get as big as they could. Our neighbors were two Basques who came from Spain, big sheep farmers. They borrowed money from the bank, but they finally went broke during the Depression. We were able to keep building up, buying a little land, renting more, expanding all of the time. We had enough brothers. Each brother had a specialty of his own that he took care of and was responsible for. Two brothers operated the sheep. One brother operated the cattle. Of course, everybody helped in between. But someone was head of each thing. A couple of the brothers were head of the farming operations. I specialized in the mechanical end of it, mostly. I went out on the field and drove as many tractors as anybody else. But somebody has to build and maintain the equipment and all that. That was my responsibility.

With more irrigation in this valley, and with the row crops finally moving in, there was less land for the stockmen to rent for pasture, because it was going into the more profitable crops. The stockmen were pushed gradually back into the hills. These two dry winters we had here a few years ago just about put us out of the cattle business. We couldn't afford to buy $100 a ton hay. So we sold out completely, in 1976, like many of our neighbors, and rented out all the range land we had.

When I was a kid there were big ranches in the county. Some have gone into estates and have been sold and broken up. During one period some of the big ranches were split by subdividers. When the Owens Valley was bought out by the Los Angeles Water District, why, a lot of Owens Valley people came into this area. They bought 40 acres, 60 acres, 80 acres. We had some real small farms in the 1920s and 1930s. They've condensed now into bigger farms again, like in the early days, but maybe not quite so big. The original ranch holdings in the 1800s were really huge, thousands and thousands of acres. But even though someone had 5,000 acres, there wasn't that much money in it, because maybe it was just dryland and rolling. It might take 10 or 15 acres to graze one steer. Later, when people were trying to make it on small ranches, history proved they couldn't do it on really small acreages, because it wasn't the right kind of terrain. They couldn't raise enough cash crops to sustain themselves.

When we bought this ranch down here in 1934 it was all dryland, no irrigation at all. Before we bought it, another group of farmers who were brothers rented it and put down five deep wells. They were going to irrigate and raise rice. It turned out to be a disaster. Four of the wells caved in, and they didn't have water, and there went the rice crop. They just didn't have the knowledge in those days how to put in good wells. The wells sanded in, and the people went broke. We happened to buy this ranch because the bank had taken it. My dad knew the bank, and they asked him one day, "Do you want to buy a ranch?" "Yes," he said. "How much do you want for it?" A rich neighbor, who is dead now, wanted to buy it too, but he didn't want to pay what the bank asked for it. So we got it, and he lost it. He was sorry later. After we were here for some years part of this land went in the irrigation district, though we had to lift the water to get it on the land.

My brothers and I at one time owned land north and south of Plainview, and to the west. We were in partnership together as a business entity, though each brother had charge of a separate operation. One brother kept the books. My parents were engaged in this too, until we retired them in 1960 out of the partnership. When we had started out buying land, the title just went under one name, J. H. Savely. Each one of us had our interest in it, but it didn't show on record. Well, that's okay for a family operation, but the day comes when the government likes to look things over. You have to sit down and get all these things recorded, what your interest is and the other guy's interest is. When our folks retired around 1960, we started sitting down. "Look, parents, you have so much property. We will pay you so much rent for it." We retired them on 1,000 acres and then bought more land on our own. They held that 1,000 acres until they died, and then it went to the sons. An arrangement had been made with the daughters in 1954 that the sons would buy out the daughters' interest in all the real estate, so at my folks' death, the daughters had already been paid off. None of them married farmers, and they weren't interested in farming — they all left home for various places. We didn't arrange to buy the property from my folks. We talked about it, but nothing was done. So we inherited it, and paid the inheritance tax, which was quite significant. But the operation went on just the same, and we paid it off. In some families it happens that the inheritance tax is the straw that breaks the camel's back. When these big tracts go into an estate, that just blows the whole thing.

We were never really into row crops, mostly because we had our setup going for years, and we were getting older. You don't adapt to changes like a young person. We just rented the land out to row croppers. Let them raise the row crops, take the risks and all. It's the new farmers with new ideas that started row crops. Row crops just gradually moved this way because they're a better income.

My brothers all had sons, and all their sons, it seems like, have stayed on the farm. We finally dissolved the partnership after 40 years, because of the next generation coming up. I have two brothers who have three sons each. Another brother has two sons. Some of them started out working for us. That's okay, but we couldn't absorb them all, and there were conflicts. We finally said, "Let's just divide up, and each brother and his sons can take land and farm on it separately. Do what they want, and go their own way." So instead of having one large unit, we now have several units. From an economic view, it would have been better to stay large, of course. When we told him we were splitting up, our banker said, "Oh my goodness, you guys will never make as much money if you split up." We said, "We know that, but it's just the times and the situation."

I was just turned 62 and I didn't have any sons who would go along with me, so I said, "What the heck, I'll retire." I have been retired for five years now. I haven't got in trouble yet. It is just surprising how busy I stay. I play a lot of golf. We travel a little. It was a little bit hard to give up all that work for about two years. You are not on the job anymore, but you wonder, "How is this going, and that going out there?" Somebody calls up and says, "Gee, we need a little help, we want to know something." Why, you just go out and help them. But that has all been phased out now. We currently own about 1,000 acres, and lease it all out. Ed here put in about 80 acres of wheat right around the house, but that's all we grow now.

Genevieve and I are now planning to transmit our property to our children by going into a gift program, so much to each child each year. We hope to reduce our estate so that on our death, the taxes won't be so great. None of the children are interested in carrying on in farming. Am I disappointed? Did you hear the story about the old Texas cowman? Well, he was a big cowman in old Texas. And he wanted his son to be a cowman. But the son said, "No dad, I don't want to be a cowman." Well, it was a big disappointment to him — but that's the way life goes sometimes.

## GENEVIEVE BRAUN SAVELY, AGE 63

I was born in 1916, near Green Valley in Yolo County. I was 20 months old when my dad had the misfortune to lose my mother in childbirth. My mother and the child both died. I have a sister about two years older, and two older brothers. My dad hired my mother's sister to raise us, since I was only a toddler and the other children were all under the age of seven. At that time my father was an orchardist in the Green Valley area — figs and I don't know what else. This aunt moved out to the ranch that he leased, but she couldn't keep it up, because she had a child of her own, and she told my dad that she no longer could raise his family for him — for pay or otherwise. Her health and her nerves prevented her. So he came to his younger sister here in Colusa County with his problem. He was broke. It had taken all the money he had to pay for our care and all. My aunt Arlene had married late in life and had no family of her own. She and my dad were close, so she said, "Bring the girls to me. My husband and I will take the girls if you can manage with your boys." So my dad moved us here. My aunt was the only mother I ever knew. My sister and I grew up in the town of Plainview and went through the local schools.

We were descendants of a pioneer family. Originally, my grandfather and his family had come to America from Germany in 1852. In 1859 he came out to California from Wisconsin. He went back and got his family in 1861, and they came to California on a boat, by way of the Isthmus of Panama. I remember being told as a child that my grandfather rode through this vast valley on horseback starting from somewhere near Vallejo. It was definitely prior to the railroad coming — which was how the town of Plainview started. The railroad is what brought most of these little communities into being.

My grandfather raised his family as a wise old German. All his sons remained in farming and became successful right in the same area with him. My dad left and went down to Yolo County at the time of his marriage, and leased down there. When that didn't work out he came back up here. My grandfather formed the W. H. Braun Company. I believe that the children drew lots so there wouldn't be dissension on exactly how he divided his property. Like all farms, some fields were better quality land than others. So he divided the property into 70 acres here and 70 acres there. The oldest drew first and then the others. Most of that property has come down through my parents' generation to my generation.

My father died ten days before my third child was born. He believed in the medicinal oil that was discovered out here in the Coast Range mountains. Some developer came in and got this oil out of the ground — they made up cosmetic and medicinal oils, face creams, hand lotions. My dad proudly brought my sister and me each a big cosmetic box with all these products in it. It was sold in the Bay Area. My dad was running the drilling derrick when he died. They had tied down a jack — a handle that pulled the drilling equipment up with a chain — and he had just turned away from it when that chain holding the handle whipped around and killed him instantly with a blow behind the ear. He was 62. My dad had lost his inheritance through the economy of the 1920s. He had a financial loss because of a rice crop failure. The man whom he had borrowed money from foreclosed on him.

My grandfather had divided his property in fives, retired, and bought a house in Plainview. My aunt inherited the house for having taken care of him until his death. She farmed her property but lived in town. She had married shortly before she took my sister and me. Her husband was weighmaster for the warehouse here. When she died, my sister and I inherited her house in town and the farming property. We were never legally adopted, but to the whole town we were Aunt Arlene's girls.

My aunt had a beautiful alfalfa field. That was another whole can of worms. When the canal came through my aunt fought it, because she had this beautiful, clean alfalfa field, and they would pay a premium price for clean alfalfa seed in those days. My sister and I used to go after school and on weekends to hand pull any mustard. When the old canal came through, they condemned the field. My aunt fought, but they condemned, and the canal went through and divided her acreage. On the north side it was plagued by seepage from then on, and the south side was only about 20 acres. Ironically, most of her days were spent battling the irrigation system that was necessary for all of the rice farming that we are now doing. It is kind of a paradox. I was old enough to remember all of this, the lawyers and all of the meetings. She wanted them to change the direction of the canal, but they had their route planned out for it.

I am proud to say I was the valedictorian of my high school class, and so was my sister. I took college prep courses but didn't go to college. After I graduated I was a part-time school secretary for the elementary principal, and then for the local doctor. I worked full time for several years. I was everything: a receptionist, a nurse, a bookkeeper, and a handy-man/chauffeur. Probably if I hadn't married, I would have eventually become an RN, because I loved it. I met a good many people that way. Once in a while someone will say, "I remember you way back." And I will say, "I remember you coming." From all over the rural hill areas, they came to Dr. Weisse. There was no other doctor.

I met Tom in town. His two brothers were my classmates. In a little community like this, the only social life was when there were school dances. Tom is five or six years older than I am, and he was out of school, but he would come back to the school dances. All of the Savelys were good dancers. He asked me for a dance, and we started dating. We went together for a long time. He lived way out in the country in those days. I hadn't even known him. It was just a name to me, the Savely name, except for the two who were in my class.

We were married in 1939. We built our house. I had saved my money, and Tom saved what little he had (his dad paid him a very minimal amount in those days — he lived at home and was provided with board and room and very little pocket money or cash). He and his brothers built this little house. When I married, I came just three miles out on this road from town. I have now lived in this house 40 years this month.

I gave up my job right away. I used to get up and go with Tom. There wasn't anything I couldn't do. We had one used car. We had no phone. We had oil lamps — the electricity was brought in later. The children came and we had to do something better than just little portable kerosene heaters. There was the kerosene stove and bottled gas. Our first improvement was a gas refrigerator. We literally had an ice box before that, ice in a box. We were milking a cow for all these children. The little orchard around here now was pasture then, for the family cow and a few skim milk pigs. I raised my own fresh vegetables. I never had any fruits or grapes or oranges like we have now, but I raised my own squash and cucumbers and tomatoes. We worked for a minimal amount — I'd say that actual cash payments for the first years of our marriage, from his folks, for his full-time work, were $500 a year.

Finally the farm partnership went from J. Savely and Sons to the partnership of brothers. Tom, at my prodding, saw what might happen if one of his folks should die. (As it is, they lived to a ripe old age.) The brothers realized that the property was all in their parents' names and what inheritance would do to the estate. So they persuaded the folks to semiretire and worked it out to a partnership arrangement. I was never consulted. It was all worked out among Tom and his brothers.

Tom was the oldest. The younger ones came into the business later. A couple of them had to do service during the war. We never left the ranch because we were already married, with a family and producing food, so Tom was never bothered by the draft, but some of his brothers were drafted. He was the sole Savely brother of the Savely and Sons then. He ran the whole operation in those war years. When the other brothers came home, they were all cut into the operation as if they had never left. That is what his folks insisted — the brothers never had to buy in. And they were given the same income that we had drawn for all the time they weren't here — over and above what they had made in the service. I was never consulted. But I didn't think it was fair.

His parents eventually gift-deeded some of their estate to Tom — but just to him, not to Tom and me. That is where I came out short on the deal. Their lawyer finally persuaded Mother and Father Savely to give some of their land to the sons in their lifetime. But instead of gift-deeding it to the boys and their wives, they gave it to the boys only. Therefore we wives don't have full ownership, as we should have. It is not community property. But the sons didn't do anything about it. Their folks were very headstrong. To get them out of the business at all, the sons had to do it their way. It has always been kind of a touchy point with me.

Now I had a separate inheritance from my aunt, which is my sole property. So my husband has some separate holdings and I have some separate holdings — besides what we own jointly together. I was the only one that really was affected by this deeding procedure because I was the first daughter-in-law. Some of the other brothers didn't marry until after the war, so some of that is legitimately their sole and separate property, which they had before they were married. But literally the Savelys had nothing but debts when I married into the family. Yet that wasn't taken into consideration. I felt that my labor and my contributions to the operation were not fully recognized. I contributed in full: I butchered hogs, I wrapped beef, I went out and turned that churn, I went right to work with Tom. I was the only daughter-in-law who seemed to help, even though at that time I had little babies. I cooked for the harvester crew. Fellows slept on cots out here behind the tank house. Tom's folks provided the food — meat, potatoes, rice, or whatever. But I grew the tomatoes and the vegetables. I had to prepare it and take it to the field. How I did it! The difference in years. I look back now and wonder how I did it, but I did. I was the only woman on the ranch. The others weren't married yet.

We married in 1939, and Philip was born at the end of that first year. Then I had Peter right away. At that time, the

Savelys were dry farming and raising hogs on the home ranch. They had leased some land, what they called the bank land, for rice away over to the east. When I had to prepare these hot meals, I would take them over to the fields. I would get up in the morning and fix breakfast for whoever was here. It went on for several years. They hired people too. In those days they didn't have the one- and two-man rice combines, they were binding the rice. I was cooking for a whole crew, and this would go on for ten days, two weeks at harvest time.

We did a lot of butchering hogs for home consumption. For hogs, we scalded and scraped the innards to make sausage, ground the meat, cut up the pork chops, canned the sausage, and put down the sausage and crocks with hot lard. All of the three sisters were married and gone before Tom and I were married, but the family still provided them with meat that we raised and prepared. Mother and Father Savely were of the old school. We used to butcher 10, 12, 15 hogs at a time. It was a workout, I remember that! My older kids remember how hard I worked. We butchered clear up to when there were movie cameras. Friends took a whole series of films of the last butchering on the ranch. It must have been the late 1940s. I always had my own chickens, for my own eggs and fryers. We didn't have Foster Fryers in my day! Tom hated chickens, because he had to do it when he was at home. But it was always my baby here. We milked a cow until the kids were big enough — they learned to milk awfully young, I will tell you. Even the girls had to learn to milk. I took care of the milk after it was milked. We had our own butter and I made good homemade cottage cheese.

Tom left here with a sack lunch that I prepared at the crack of dawn, and I never saw him until 8:30, 9:30 at night. The kids would be in bed and asleep before their dad ever came home. He raised the first four without ever really knowing them, except for occasional Sundays in those days. A lot of Sundays he worked also — especially putting in and taking off the crop. Sunday was just another day. By the time he quit working so long, the first four kids were up and away to college and in the service and all. The younger two had his attention, the circumstances were different. We both worked very hard. We didn't have electric pumps to life the water for the rice, they were gasoline. We would get up here at an unearthly hour, and Tom would take the one little home-built truck — he had taken a little Chevrolet and put a little box in back to make it into a pickup — he didn't even have a standard pickup then. Those pumps, about five or six miles away, had to be serviced night and morning. He would go over and oil and put gas in the tanks while I packed the lunch, then come home for breakfast and pick up that lunch and leave. When he came home at night he ate supper. With lanterns, they went and serviced the pumps to go through the night. He had something like 12- and 14-hour days. Of course, I did too — of necessity.

I never did any farm work. It would be pretty hard to drive a rice truck when you have four little kids. I had to get them to catechism, and to doctor appointments, and orthodontistry, and all that. I probably would have done farm things if I hadn't had my time fully occupied. It was physically impossible. I have always gone for parts and things, though. I would get a call from the home ranch, and run down to Oakfield or up to Colusa to pick up this or that — a lot of errands. As we got on our feet after the war, I had a car. For a long time there was just the one car that went to work with Tom. After the war we bought a second car to save the children the horrible long bus ride — the school bus came to pick them up, but they had to go all over Timbuktu and were completely exhausted by the time they got to school. I would pick them up at school and I would do my own shopping then. But for years I never went anywhere except to church. When the kids were little, Tom and I used to take turns. One of us would stay with the little ones one Sunday, and the other would go to Mass. The following Sunday, vice versa. When they got a little older, we would drop the tiny ones with my aunt. She delighted in looking after them that hour. We could go to Mass and pick up the children before she walked to her Methodist Church. That was the only recreation I had for years and years.

When Roberta was nine months old I folded up with complete exhaustion after having her and then resuming the regular chores. By that time there was all kinds of extra activity. My husband was one of those dedicated farmers, from dawn to dark. I just plain folded, because Roberta was the fifth child. That was the first time I ever had any hired help at all.

About the mid-1950s we no longer milked the cow, because after a while there wasn't the need for all that milk and cream. Tom didn't want to have to go out and milk, after he put in those long days. With the machinery the brothers were buying, they would go out for hire. When they had completed their own harvest, they would help out somebody who didn't have their crop in. With rice you are fighting rains like today. The neighbors would pay a premium price for Tom and that harvester and his brother and another harvester to take the crop off. All that custom income went into the pot of the Savely Brothers.

When they dissolved the partnership five years ago, I assumed that I would have full ownership of our property under community law. But I didn't even know some things until they dissolved. The tax lawyer said, "These parcels are deeded to the sons only in this gift program." The brothers were a tight entity. Out there they would have their meetings at lunchtime with their mother. Tom is a very noncommunicative person. I was just not told things. When he retired the brothers met and met. Finally they agreed. I had assumed that under California law, our property would be a full 50-50, but I found out that legally it wasn't really so. I make no bones about it: It is a very sore point. As long as he was operating as part of the Savely Brothers, he brought home his full share of the partnership income, but settling the estate changed things. With the inflated valuation of this row-crop land, the estate is now very large, and it poses a tax problem. I didn't realize how vulnerable I was. It wouldn't have mattered, perhaps — but the first of March last year, Tom piled up an airplane out here. He wasn't scratched, but he could have been killed. And I would have been thrown into a situation where I could have been wiped out financially. I didn't know before because I didn't have to pay any attention. But when something like that happens, you start looking at these things. I didn't realize what a small percentage of ownership I actually had — even in this field around here.

Tom's folks were in a gift program in their last year. They deeded land to Tom but it actually belonged to both of us, because both of us had to sign when we borrowed money to buy this ranch. The banker made it very clear. You see, my aunt had died 21 years ago and left property to me. The banker came here because I was too busy to go to the bank. The brothers took care of all the loan, and then they came down here and said, "Genevieve, you sign this loan." The banker knew that I didn't know a lot about their business, because everybody knew how close-mouthed they were as a family. The banker came of an evening with these loan papers and he said, "Genevieve, I want you to understand this. I know you own property in town, and you own a farm acreage, and I want you to understand that when you sign this loan, all that is subject to being sucked in to the Savely operation." Well, I didn't have much choice. I was just offered the paper where you sign it. Nobody took into consideration that I was the only daughter-in-law with sole and separate property of her own. The rest of them didn't have anything, except what they got in their marriage. The law is that when you borrow money, the bank requires the wife's signature, so she knows what is happening. The Savely Brothers couldn't borrow without their wives' signatures. But the banker knew I had property, and he wanted me to know that if they didn't meet the payments, the bank could come in because I had signed the loan paper and take my free and separate property because my name was on that loan. The other daughters-in-law didn't have any property, so there was no skin off their noses if they went under.

The banker looked after my interest. But everybody was standing there, and they didn't expect me to object. "Wait a minute. I don't feel very enthusiastic about this. What protection is there? Is there any codicil that could protect me?" The banker was a friend of mine and knew that I probably hadn't been informed. I was never even told they were buying it until they showed up one day — "Here, sign this." That is the way they operated. "Sign here, sign there" — without ever even saying beforehand that they were considering buying. None of us wives knew what was going on, except what they chose to tell us. They would just come home with so much partnership income.

The fellows met at the home place. When they went to individual ownership, we wives were told to stay out of it. "We are looking after your interest." We weren't even allowed to make suggestions — they made the decisions. It was the old school — the way they were treated by their parents is the way they treated us. We just had to trust that we would come out all right.

Oh, we have seen lots and lots of changes since we farmed. We have seen all the fences go down. It used to be, you looked out here and there would be a cattleman moving his cows, or a sheepman moving his ewes and lambs. There was always stock being moved. It used to be sociable, because it would be your neighbors. Maybe they would come in for a cup of coffee — they would keep an eye on the fence and the sheep would kind of move on down the road and the men would come in — they were welcome. It would be wet and cold like this. Maybe they would be moving their ewes because the water was rising, out of a field, to a field a little higher and closer to the foothills. Now livestock is trucked. You don't see any on the roads. With the advent of row crops, they make what they call a pad road where the fencerow used to be. They farm every inch of it up to that — the pad road around the property is for the irrigator, so he can hop out of his pickup and go out on the check to irrigate and to drain or whatever.

Over the years, from the days of very hard work and very little return, things have gotten steadily better. We are really quite comfortable now. Farmers finally started making a little money, over and above their payments, and got a little ahead, either to acquire more land or make investments to help out for retirement years. That was the postwar boom. With the improvements in machinery, farmers could harvest with less expense. There were world markets for rice. But in those early days we just worked for an existence, or I guess what you would call the good life, the healthy life. They were long, hard hours.

The older four of our children used to resent it that to some of the kids who lived in town they were "one of those Savelys." The family was known because there were so many of them. They were a large farming operation. When you really get down to it, per acre, with that many it wasn't that big of an operation, but it seemed so to people in town.

Our kids never did help their father with farming much, though. There wasn't enough outside work. They worked away from home, as they got up into high school. Philip worked for a land leveler, drove engines, heavy equipment. Pete, until he went away to the university, worked for a neighbor down the road. He got up at 4 o'clock in the morning to cut hay, irrigate, and bale hay. He worked for others in almonds. But there was no room in the family operation. What work they did hire went to Tom's nephews who lived close to the home place -- there were too many nieces and nephews.

My children have chosen not to stay in farming because they remember the long hours. One son-in-law, Carol's husband, would love to farm, but Carol won't let him. When they were first married he had a fine engineering job and he was sent overseas. They had work/recreation weekends paid for by the company, and the company moved him over and moved him back. She knew you have to save money on the farm. There are no insurance plans, no fringe benefits. He thinks farm life is great, but Carol remembers that it wasn't always like it is now. He only sees it more or less in the good times. But Carol is in her 30s and remembers that it didn't come easy. Phil remembered how hard he worked for the adjoining ranch neighbor. He would get sinus from cutting that hay till he couldn't breathe for days after. That lovely smelling new-mown hay! Then Ed, after he got into high school, worked for a big farmer here in the summertime, running a harvester and all that. He was known to be a darn good mechanic, like his dad. He was drawing a man's wages at another farm because he is responsible, reliable. But he too would get all plugged up with sinus from the dust and all. In putting in this crop of wheat around here this year, he would come in: "That is the most boring thing I have ever done in my life, sitting on that engine." He put on these earphones with the radio in them, but was still bored to tears going around and around. His thing is taking the aircraft engine out of that airplane out here, completely overhauling it and reassembling it, painting, building up, and covering the wing. He likes the challenge. Now that would drive me up the wall. But he has the patience and the ability. Ed takes all the aircraft journals, he takes Popular Science, and keeps up with the new things in machinery.

Our second son is in Hillside now. He had a mental breakdown over an unwanted divorce. He met and married a girl in Pennsylvania after being in the Atlantic fleet all his service years. He went to work in the steel mills in Pennsylvania. He was married for four years, then out of the blue one day his wife decided that she wanted a divorce. It came as such a blow, and he didn't know how to cope with it. He told the judge: "Marriage is for life." We went back to Pennsylvania. We said, "If she is through, she is through, and there is nothing you can do about it, you can't fight it." He said, "I will just sit here. I will go to jail. I will do anything. I am not giving her the divorce." He just couldn't accept it. So we hired a lawyer, brought him home here, and took him for counseling. The lawyer sent word that she was remarried before the divorce settlement was ever through. It just hit him too hard. So he is in a rehabilitation program. We are still having trouble with him, but he is much, much better.

Tom grew up in farming and that is the only life he ever knew. But the boys have been in the service and have worked away from home. The girls went away to college and married, and their husbands are doing other things. The world is much broader for our children than it was for us. They like the culture of a bigger city. They come home holidays and they see their friends in Plainview, who grew up, got married, and stayed right here in Colusa County. All their cousins have just one idea — to become successful farmers. There is nothing for the young people here but a few bars and a bowling alley to go to on a Saturday night. I'm glad that my children are choosing to do other things. Tom makes no bones about it, though. "That's all I ever wanted to do. I just can't understand it." And he really can't. Most young men would be green with envy at the opportunity to get into farming like they could. Ed will be 22 next month. He could be a wealthy young man in a couple of years, but money doesn't mean that much to him. He would rather enjoy doing what he is doing. The others were allowed to make their choices, and they chose not to go into farming. Philip came out of the Air Force and got his Master's in Journalism. King Broadcasting offered him a job in Spokane and later he was offered a job in Portland. He has been a legislative reporter for the last two or three sessions of the Oregon legislature. He just loves it and he is very successful. You can see him on NBC here every once in a while.

    Roberta is a tall, pretty young woman with long dark curly hair, big hazel eyes, and a white smile. She is dressed somewhat modishly, but one could equally well imagine her outdoors in blue jeans. In conversation she is open and expressive, a quality of thoughtfulness pervading every response. In her mother's presence she is deferential and sympathetic.

    Her younger brother Ed has glasses and a beard. Slow to speak, not very articulate, it is hard to get him to talk even in private. It is not clear what lies beyond his quietness, but he does not seem to possess the passion for life of his sister.

## ROBERTA SAVELY, AGE 26

I was born in 1952 and grew up near Plainview, population less than 2,000. I had four older siblings, with a gap between them and me. Then there was another gap, and four years later my younger brother was born. When I was nine or so my brothers and sisters started to leave home, and I did not have anyone to play with because I was living out in the country. Eddy was still little and his interests were not mine. I learned how to entertain myself. I would read and play dress up. I had a great imagination. That is one thing I am thankful for, I am never bored. I can always find something to do.

I didn't have regular chores to do as a child, but I remember picking worms off the tomato vines, gathering the eggs, feeding the chickens. There were no "have to's" but contributions were appreciated. I did help my father occasionally. All the kids would go with him on Saturdays at some point or other, maybe to help him out in the shop, hand him tools. When he surveyed the land, I was always his feet for him — I would walk with him and take the stake for him to do the sightings. All of us learned to drive at a very early age, too, in order to run errands. If he had to move equipment from one spot to another, we followed behind in the pickup, for the return trip. We all learned to drive stick shift, as soon as our legs could reach the clutch. I did tend to be interested in outdoor activities more than housework because of my size. I am a large person, and I tended to do more masculine things. I don't remember playing with dolls much — I would be playing with Tonka Toys. I wasn't real strong on the domestic, feminine things.

In high school I was exposed to music and drama. Music was my social ticket — that was a group activity. I was an instrumental performer and a singer both. I wasn't in 4-H at all. Mom had ulcers or something. She was pretty busy and didn't have the time or energy to chauffeur me, whereas the older kids had done it together, and her efforts then were worthwhile.

I graduated from high school in 1970. Dad expressed the opinion that he would not like to see me go to Berkeley, but to a college more stable and less chaotic. I was fortunate that we could afford Santa Clara University. I had heard that the Jesuits were very educated priests, and I was terribly curious about comparative religions, so I went there. It was a small campus and probably an easier adjustment to go from Plainview to Santa Clara than to Berkeley. I was a sociology major, and I wrote for the newspaper. I stayed there two and a half years.

I left Santa Clara for different reasons. I had an internship in the Veterans Administration Hospital with mental patients. I was interested in social sciences — human behavior and journalism combined. After this internship I was overwhelmed with the fact that for these older men, change was so hard in later years. If you wanted to change human behavior, you had to do it younger. I had always worked with kids. I did lots of baby-sitting besides teaching swimming at the local pool. I began to be interested in child development. The 1960s were a factor in my college education — I had some idealism. I was ready to go to a more diversified student body, so I transferred to Davis. I liked the idea of being in the valley again. And by then I was tutoring in a migrant camp, and the Santa Clara campus was pretty wealthy. The extremes really got to me. There were sons and daughters of professional people from urban areas who would have excellent parochial educations. (When I first heard the name "Aristotle," I thought it was a disease! I was real embarrassed because I didn't have the same background.) Santa Clara was a rather elitist college and after a while it rubbed me the wrong way. Davis seemed more comfortable. There were some agricultural people there, people more my type, who weren't so worldly — yet were interested in all sorts of different things. At Santa Clara I was wearing my bib overalls when most of the girls showed up in their I. Magnin clothes. There wasn't anybody else in bib overalls at Santa Clara — though there were plenty at Davis.

I went straight through the child development program at Davis and graduated in 1974. By then I had changed from journalism as a career goal to working with kids — but in what capacity, I wasn't sure. I did volunteer work with terminally ill children in a play therapy program. You'd give the children a syringe and a stethoscope, and through their play they could take out their aggressions, their frustrations about their experience, the feelings they had toward the doctor.

The summer between my junior and senior year I went to Israel for a month with a health clinic. We worked mainly in the Kibbutzim in the countryside, the collective farms. That interested me a great deal. Then a job came up in France for the fall. A French family wanted an American girl to come live with them and teach their children English. I though why not? So I went and stayed for two years. They call it <u>jeune fille au pair</u> — it's a live-in baby-sitter and housekeeper position. I taught English to the children along with my other duties. I became part of their family and lived the French life-style. There were six children, aged 4 through 14. They were a very Americanized French family, the husband worked for IBM. I learned French from scratch. During the second year I stayed with a family that spoke no English so I accelerated in my use of French then.

At the end of the two years I decided to come back to the United States. My work permit was not going to last any longer, and I had an uncle who was very ill, and I wanted to be here. It was time to come home. I lived with Mom and Dad for about five months, and took some classes at the local community college in business, and volunteered in a hospital. Then I became a preschool teacher for about two years in Hillside. After that I moved down to the Bay Area. I am job hunting right now, debating what to do next.

There is no place for me here in Colusa County, so I don't expect to be returning. I define education as exposure. I have been very lucky that through my formal education and traveling I have been exposed to so many different things. The first trip we took was summer of 1971, to visit my sister in Germany. Mom wanted to see her grandkids. We went over for a month and had a great time. We learned about Europe — geography, history, and fine arts. I was thrilled with the experience. When we came back to Plainview I wanted to talk to somebody about it, but I was totally cut off. When I started bubbling over about this trip, a friend in my high school class said, "What are you trying to do? Impress me with how rich and smart you are, to go traveling all around the world?" That just wiped me out. There was nobody to share it with.

That experience sums up my dissatisfaction with the life-style in this county. If you are interested in classical music, there is no group that studies classical music. You do it on your own. There are no groups in the whole county that you can share cultural interests with. Probably that is because the population is so small. They are not purposefully shut off to learning, they just don't have exposure to things. How could they be interested in classical music if they have never heard any? There are no concerts, no theater, nothing. Not even quality Hollywood or foreign films. Maybe I wouldn't have missed that if I hadn't been exposed to it somewhere else. Catch 22! It rings true that it's hard to keep people down on the farm after they've seen Paris.

People know everything about you here, in the sense that they know what car you drive and the name of your dog — but they don't really know you and your feelings. You run into people you know every day, but you are not open to sharing some things. I am finding that is what I value most in this world: learning about feelings and sharing ideas. But it has to be reciprocated. It's a sad thing — people here in my age group hang around the bars, go water skiing. They just now sold the bowling alley to a couple of truckers who wanted to get off the road, so maybe there will be snacks and bowling for after-hours or a rainy day. People tend to marry early here. The successful young farmers save up, and when they are not putting that crop in, or taking it off, they see every change of the show at Lake Tahoe. They go up for the dinner show and gamble a little over the weekend. Some of them have homes up there, or rent, and go winter and summer. But the young people here who do not have the means have absolutely nothing to do. The county is still focused on agriculture, and agriculture only.

The minority question here is very intriguing. There are those Mexican families who choose to live in town and really become part of the community, and their children marry whoever they want — that can be used as an indicator of social flexibility, I guess — and nobody thinks anything of it. They have been here a while and they work regularly. They have acclimated, they are even volunteer firemen. If they make an attempt to assimilate, they are accepted. The migrant worker is another category. There is a hierarchy even within the Mexican community. New migrants are on the low end of the scale, and the more settled workers, depending on how long they have been here, are higher up in the pecking order.

At one point I thought, "I have gotten this education now, and it is my responsibility to return my education to my community." I was taught that by the Chicano movement and the Black Power movement. The kids were to come back and contribute to their surroundings, to try to improve things. I thought about administering day care centers in rural areas. That would free the women to study in community colleges and expand their horizons, and thus help them get more involved in organizations and maybe in political activities. But I am not ready to do that, because I am single. I wouldn't want to come here as a single person. You have to make choices, that is the point. I have a need to be exposed to a lot of different things.

At one time in the late 1950s my oldest brother expressed an interest in farming. But at that time my dad and his brothers were building up the operation and there wasn't any room for Phil. He was the first one of the extended family to go away to college. He did a lot of speech-making for the FFA. A college professor told him that he had a great voice, he was very eloquent, and he should go into radio and TV. He learned there might be an alternative profession for him. So he studied broadcasting and that became his career. After that he never considered going back into farming. And my second brother was never really interested in it.

Ed trained under Dad. He learned all his mechanical skills as an apprentice to him. He also went away to school and gained new ideas. He is very interested in appropriate technology. If he were to come back here he would like to implement solar-energy rice dryers, for example. But he isn't finding a great open-minded attitude to implement those changes. It frustrates him.

With all three of my brothers, it was hard to work with my father. He knows his business well and he is demanding. He is a real perfectionist. I remember helping him with walnuts. He wanted me to help sew up the sacks with a big hooked needle, an awl. He could whip up those sacks in three seconds flat. He expected me to be able to do it as fast as he could, the first time!

I thought at one time about learning how to be an orchardist, I thought I could handle that. It was only a fleeting thought. But farming has never been considered to be a very viable option for a girl. Women were never really part of the operation in our family. In some families women will occasionally drive tractors. I learned how but there were enough men to do it. I think, more importantly, we didn't get the feeling we would be part of the decision making. Our physical labor might be utilized, but not judgment on what crops to plant and when. My mother was depended on to run errands, to go get parts and deliver items out in the field when needed, and the domestic land right around the house was her territory (the garden, the chickens, and that). But she wasn't included on the big decisions. That was a source of frustration. She had had experience in the working world, but when she tried to contribute that, it was discounted. She would have been interested in contributing more. But my father and my uncles were a clan, in all respects of the word. It was an anachronistic kind of governing process.

Recently we had a family meeting among my siblings about how we would manage the family property. We are very dedicated to cooperating and keeping the holdings in common, never losing the perspective that it is a business, and the right hand needs to know what the left hand is doing. We will have open meetings and talk about how things should be run. In our immediate family we are really dedicated to an orderly transition, because we have seen results when that collaboration doesn't take place.

When I come home now, I don't make much of an effort to see those classmates of mine who have stayed here. Most of them married right out of high school and already have children who are six and seven years old. I have just gone in a different direction.

## ED SAVELY, AGE 22

I was born in 1957. I was the last one in the family, so I was here almost by myself as far as having brothers and sisters around. Roberta was the only other one, and she's almost five years older.

I was about average as a kid, not much different from my school friends in town. I didn't have any plans to go to college until I got interested in aviation. Then a colonel in the Air

Force Reserve started an Air Scout group — he was a farm manager for a big outfit in the county, so he had regular hours. He passed on his interest and knowledge about aviation to us. We went to Oakfield weekly for this Air Scout group, and he gave us ground school. It didn't cost us a thing. It was all weather and aerodynamics, gravity and depth perception and all. In real ground school, you also have to pass a written test. Then you can take flying lessons. I have always been interested in airplanes, and in the Air Scout program I got interested in the mechanical aspects of aviation. I have always been around machinery, since my dad is a mechanic. I always worked with him out at the ranch, doing repair and construction.

I heard about a vocational program in Hillside, so I decided to go down to college. I spent the first year and a half in a boarding house. Then Roberta and I shared an apartment for five months. I was going to Hillside City College in a two-year course. I stayed a third year, so after I got a vocational certificate in Aviation Maintenance I got my AA degree in 1978. Since then I have been doing a little work on airplanes around here — my own and dad's airplane. I put in this wheat crop this fall in the field surrounding the house too.

Actually I don't want to stay in this county. I would rather go to the Northwest — Washington or Oregon. I like rainy, cool weather. I don't like the $110^o$ temperatures here in the summertime. I guess I could stay here and be a farmer. I would be all right at it, and I have an excellent opportunity here. But I don't have that much interest. I like some things about it — living out in the country is one of them. I am not looking forward to moving to a large city. But I have to try it. Mainly it is just the situation I am in here that makes me want to do something different — being in competition with all my relatives. Maybe I am afraid of being intimidated.

The other day Dad was telling me that when he got out of high school there was no other opportunity for him. His brother-in-law got him a job as a mechanic over in a garage in Colusa. He lasted about one week there; he didn't like it. Then he got a job for about a year and a half working for one of those farmers out there where he lived. That was what he liked. But he didn't like working for somebody else.

When I was in high school I did some work for my uncles. I really didn't like working for any of them, so I went to other farmers and got jobs and worked totally out of the family. And I did enjoy that. I don't mind any of the tasks in farming — driving engines and what not. But when stuff breaks down, I like working in the shop fixing it more than I like sitting on a harvester or a Caterpillar all day. Mechanics to me is like putting a puzzle together.

You do what you like, and what you have to do.

# Part III

And he gave it for his opinion . . . that whoever could make two ears of corn, or two blades of grass, to grow upon a spot of ground where only one grew before, would deserve better of mankind, and do more essential service to his country, than the whole race of politicians put together.

>Jonathan Swift,
>Gulliver's Travels:
>Voyage to Brobdingnag

All that shrieking, bellowing machinery, all that gigantic organism, all the months of labor, the plowing, the planting, the prayers for rain, the years of preparation, the heartaches, the anxiety, the foresight, all the whole business of the ranch, the work of horses, of steam, of men and boys, looked to this spot — the grain chute from the harvester into the sacks. Its volume was the index of failure or success, of riches or poverty. And at this point, the labor of the rancher ended. Here, at the lip of the chute, he parted company with his grain, and from here the wheat streamed forth to feed the world.

>Frank Norris,
>The Octopus:
>A Story of California

In the 1920s fruit packing sheds were common in Stanislaus County. Reprinted with permission of the Department of Special Collections, University of California, Davis, Shields Library.

# THE SPREADING URBAN WORLD
## Stanislaus County

In 1908, U.S. soils surveyors submitted the following observations of Stanislaus County: "A rapid and marked change in the character of the population after eight years of water is taking place . . . a change from a scattered population of grain growers to a dense population of dairymen, vineyardists, orchardists, truck and melon growers . . . from a 1,000 acre farm unit to one of 20 acres."

The earliest public-project irrigation in California was implemented in Stanislaus County through the Modesto and Turlock irrigation districts in the 1890s. In 1901, when water flowed down through the Turlock ditches for the first time, a revolution in the region's agriculture began. The good soils, moderate climate, and mostly gentle topography of the area presented great possibilities for a diversity of crops. The great grain ranches of the nineteenth century were broken up as tracts of 20- and 40-acre parcels were developed, many of them in "colonies." New towns sprang up, and the number of farms in the county increased from 2,687 in 1910 to nearly 6,500 in 1935.

Under irrigation, a farm of 20 to 40 acres was quite adequate for a family dairy, orchard, or truck farm in the first decades of the century. Throughout the 1920s small farms flourished. Swedish and Portuguese immigrants ran many successful family dairies of a few cows. The Turlock region became the "Watermelon Capital of the World," shipping watermelons and cantaloupes all over the country. Vineyards and orchards proliferated along country roads. Small poultry growers produced hundreds of thousands of chickens, while the present prosperous turkey industry got its start from an ingenious entrepreneur who fixed up some insulated railroad refrigerator cars as a turkey poult hatchery.

In 1922 a little book published by the Interchurch World Movement, called Irrigation and Religion, surveyed Stanislaus County and linked church and community activity with agricultural prosperity brought about by population growth as large ranches were subdivided. While irrigated farms were the basis for the healthy regional society praised in the book, these were also the days for formation of farmer cooperatives and various food processing facilities. This period was truly the golden era for a small farmer, and Stanislaus County was a good place to be.

By 1938 nearly 90 percent of the county was under some kind of cultivation. The average size farm was 93 acres, with 59 percent of all farms in the county from 10 to 49 acres. Nearly 80 different farm commodities were produced, many of them specialty crops like almonds or apricots. By the early 1940s Stanislaus was growing more peaches than any other comparable place in the world.

The Depression years, however, took their toll, and some farms in Stanislaus, as elsewhere, were foreclosed as farmers were unable to meet their financial commitments. Compared with the boom years of the 1920s, the 1930s were an era of persistence and adjustment.

After World War II, developments in agricultural science and technology had the ironic double effect of increasing farm productivity while making it harder for small operators to make ends meet. The high costs of new equipment and new technologies put economic pressures on small farms everywhere. The trend to small units in Stanislaus County, made possible by irrigation a half century before, was reversed. A 40-acre orchard or a 20-acre dairy was no longer adequate for a full-time family enterprise. Former flocks of a few dozen laying hens scratching about chickenyards became concentrated in huge henhouses containing 50,000 birds in cages. The county became a national leader in turkey processing, with several large companies taking turkey poults from hatchery through to packing and retailing.

While production yield and quality zoomed to new heights in nearly every commodity, the farm population began to fall. By the 1970s Stanislaus farms had declined by nearly one-third from their all-time high in the 1930s. Small farms became larger as their owners found it necessary to expand just to keep up. Individual 20-acre parcels began disappearing as neighbors absorbed them or outside investors brought them up.

Nevertheless, compared with other California areas, Stanislaus County is still characterized by relatively small family farms. The climate, soils, and water are right for farming intensively, and a strong infrastructure supports agriculture. In 1980 Stanislaus ranked high for production of almonds, apricots, dry and snap beans, boysenberries, melons, peaches, peas, spinach, sweet potatoes, walnuts, and dairy and poultry products.

A new kind of economic growth is also evident. Located in the heart of California, the county has been an attractive site for industrial and residential development. Canneries and other food-processing facilities have seen particular expansion. Modesto's population has been one of the fastest growing in the state, and the city boasts numerous new shopping centers and subdivisions. Rampant county growth has led to considerable controversy and negotiation over land use issues.

In the old section of Modesto a 1912 memorial arch spells out "Water-Wealth-Contentment-Health." It stands as a reminder of the county's ebullient days of enthusiasm over the new prosperity brought by irrigation. Stanislaus County today, however, faces new problems and new conflicts over use of resources. Farmers here are always conscious of their neighbors, for there are many of them. Agriculture, though still a major underpinning of the economy, is only one of many activities in a bustling county.

# PORTER WOODS
## COUNTY AGENT

Stanislaus County has a large, bustling Extension office housing nearly a dozen farm and home advisors. Working out of the county seat of Modesto, they range widely in a farming area that is richly diverse and full of history. History is being made now too, as the county struggles to preserve its agricultural identity under the influx of enthusiastic new settlers swelling the population in their search for the good life.

Porter Woods has been in Modesto for 15 years and loves it. He is a warm man, graying at the temples, with kind brown eyes and a mellow voice. He speaks of family farms and the international monetary system with equal concern, and tries to make connections where he can.

Stanislaus County lies in the San Joaquin Valley between the Coast Range and the Sierra Nevada. It's about 60 miles wide, and from the north to south it's about 25 miles. The trough of the county is the San Joaquin River to the west. The west side of the county is made up of soils transported from the Coast Range, mostly sandstone and shales, that have weathered into loams and clay loam soils. On the east side the soils developed from materials from the Sierra Nevada, and they tend to be more sandy soils. Just east of the trough of the river are some real heavy textured clays. This is also an area where salt has accumulated, so that land is used for pasture, for livestock, a strip of about five miles wide. East from there come more of the field crops and the deep-rooted crops, the trees and vines, clear up into the foothills. Then we find heavy soils again and irrigated pasture. Some rice is grown in the county, but not very much. Beyond that, into the rolling foothills, it's mostly rangeland. How productive it is depends on rainfall.

Practically all the county is served by irrigation districts. The west side has the Delta-Mendota canal, and a little water from the California Aqueduct. The east side has three irrigation districts, the first districts organized in California, going back almost to the gold mining days. Practically ever since white settlement began, there's been some irrigation, and of course the irrigated areas have increased since then.

Farming in this county started first with dryland grain and then went to irrigated grain in recent years. Generally speaking, there's been a continuous trend from growing field crops to the various tree crops — peaches, apricots, almonds and walnuts, and grapes. Farming here has continually moved toward more intensive agriculture, and higher-income crops.

Our livestock industry is primarily a cow-calf operation, mostly on rangeland pasture. There is one small feedlot and one large feedlot. Dairying has been very strong here. Back during the 1950s there were about 1,600 dairies. Today there are only about 500, mostly Grade A dairies. Those that were lost were the small dairies and the Grade Bs. Dairies are fewer in number but larger in size. The large commercial dairy now has about 200 head of milking cows, with some up around 500 head.

At one time there were many farms raising poultry — real small operations. But economics has squeezed all those out. We have just a few farms producing fryers, and one huge operation distributing to stores all over California. Stanislaus County has an egg-producing industry here too — maybe 50 operations. It is highly integrated; many of the growers are financed by the feed suppliers, and many of them market directly. There is no longer the small operator who produces a flock of fryers, or some eggs, and then puts them up for sale. The poultry man already has a home for everything that he's producing, on a contract basis, and his feed is supplied.

Back in 1959 there were 6,000 farms in this county. The 1974 census shows 4,500. Meanwhile the average size has gone from 140 to 165 acres. Most of the land is developed now, or it's in hands that are going to keep it. Some land has gone out of production because of urban development. When I came here in 1965, Modesto was still a sleepy community with some development along country roads. Farmers were selling off little parcels, maybe five or ten acres. We're located relatively close to San Francisco, and we're close to the mountains and resort areas. People coming from back East wanted to retire in this area, so there was a big demand for five- and ten-acre plots. And there still is.

Parceling has been a real problem. About 1969 I became very concerned about what was happening. There were no restrictions on the urbanization of land and the splitting of farms. I put on a couple of workshops to make people realize what was happening. I could foresee a sprawling Modesto with long fingers, hopscotching into agricultural lands, with all kinds of problems. I didn't want to happen here what had happened in San Jose and Santa Clara counties. Our workshop was called "Stanislaus County — the San Jose of 1980." It really stimulated and rallied the citizenry of the community. Out of all the turmoil that took place over four or five years, the county agreed it would no longer allow residential developments in outlying areas. They had to be on sewer systems, which meant they had to be incorporated into the city. The city agreed not to allow sewer trunk extensions, because a lot of vacant land within the current sewer service area still was not developed. This restriction has forced development of that land. Of course, all the time the developers and the real estate people, at least some of them, were screaming for extensions. Now each year the City Council has a growth committee that takes into consideration the extension of sewer trunks and the bringing of more land into the city. When the county general plan was agreed upon, all the towns had the option of designating what they called "urban reserve" — an area that included agricultural land but ultimately could be urbanized as the need arose. That's still a bone of contention, because developers want to keep on moving out.

Stanislaus is one of five or six counties in California that have made the greatest progress in land use planning. We think growth should be orderly and planned, so that it doesn't destroy a lot of farmland as the cities grow.

Parceling of farmland had become a problem for other reasons. Say a man had 40 acres and had the bright idea that if he had four children, eventually he would want to split it up into four ten-acre plots. That's fine for a few years; then it ends up that ten acres is an uneconomical unit and no farmer wants to farm it. Because of inflation and the economic situation, then the ten acres would be abandoned and go to weeds (maybe a few livestock). It would essentially be a loss to agriculture. Today there are definite restrictions on parceling, although the minimum parcel size in agricultural zoned lands is still ten acres.

I am on a task force committee that is addressing this issue. Our challenge is to provide for people who want to live on small acreages. Some areas are adaptable to that type of urbanization. Where there already is parceling of land, we could designate that as rural-residential and allow small splits. But on our bona-fide agricultural land we should upgrade the minimum to 20 acres, 40 acres, 80 acres, or even 160 acres. There are farmers who realize that they may have to accept that kind of zoning. But there are always others, the last of the rugged

individuals, who say, "Damn it, if I own a piece of land I want to be able to do anything I want with it." In today's society that doesn't work anymore, because anything you do will have an impact on others around you. I would like to see the preservation of agriculture on the good soils just north of Modesto — let the town grow out to the east where the more limited heavier soils are. There are fewer options for that kind of land. Much of it is just pasture. Some parceling will continue along the bluffs of the river.

Stanislaus County in the last ten years has experienced 20 percent population growth, one of the highest in the state. Even the Chamber of Commerce has been surprised. There are many factors. It's a desirable area. Modesto is a good clean community and close to the country — you can see the orchards and the farming. People like that. Many retired people live here and many part-time farmers can supplement their incomes by working in industry. We have a large processing industry — canneries, wineries, dehydrators, freezer concerns doing vegetables. We have International Paper here, making milk cartons for the large milk industry, and other supply industries, like farm machinery and plastics. The existence of job opportunities in these industries has provided a means whereby part-time farmers can get by, or they can pursue farming as a hobby. Some part-time farmers work for other farmers. Many women work in the canneries during harvest season. They're perfectly happy to work three months a year and make good wages.

Farm statistics give some idea of trends in this county. In 1959 there were about 1,300 farms of ten acres or less. In 1964 that figure dropped down to 780, but in 1969 it jumped back up to 913, and in 1974 to 969. I dare say it's higher than that today. Now a ten-acre plot is certainly not an economic unit, a far cry from it. It's a mode of living — that's really what it boils down to. Still, a good number of the ten-acre farms are producing something commercially. Maybe they're growing their own beef, or a few sheep. Some people object to calling a ten-acre plot a farm — but who's to say what's a farm and what isn't? The census definition is ten acres and $2,500 in sales.

There's always a demand for land. Because of the big investment in machinery and equipment, as capital has replaced labor, a farmer wants to increase the size of his operation. It doesn't necessarily have to be adjacent to his operation; it could be a mile or even farther away. Average land values in the county over a period of 20 years have gone up tremendously. When I came here I had the opportunity to buy a piece of land at $500 an acre. It was open land, but very good land. Today it's in almonds and it's probably valued at around $7,000 an acre. I don't think you could touch an open piece of land today for less than $3,000. Anything that's developed — almonds, walnuts — is of even greater value. Those very high land values mean that entry to agriculture today is limited to those who are well heeled, or else they have to buy small acreages. The opportunity for newcomers to agriculture is very limited, not only because of the land situation but also because of the high cost of machinery and equipment. A young person who wants to get into farming does it through inheritance or through other employment, gradually paying for land over a number of years but working elsewhere to make a living.

Economic pressures that have hit other industries before are now hitting agriculture. In Henry Ford's day he had the ingenuity to invent a car and then start producing it. He got financial backing and finally made a mint, became independent, and became "aristocracy." Maybe we're just in the beginning stages of that in agriculture, where the really ingenious producers will come to dominate the industry. I'm not necessarily justifying it, however, because I think agriculture and people being on the soil is kind of a basic thing. It's part of our make-up to want open space around us, and maybe dig our toes in the soil.

There has been a kind of revolution, I think, in consciousness. For years and years, people wanted to get away from the farm. If they found a good way of making a living, they could get out of the "boonies" and into the city. Now there seems to be a reversal of that. Migration out of the cities is one sign of it. But there's a lot of frustration among young people as to whether they can really ever get into farming.

Recently I spent some time in the Middle East, and I was struck with the contrast between their standard of living and ours. When I saw the poverty in Egypt it really brought home how affluent we Americans are. But if we continue to have the same economic pressures on us — inflation and deficit spending — the same thing is going to happen right here, with population growing the way it is. I see an increasing disparity between the ends of the economic spectrum in this county. I don't care what kind of government programs there are — social security or unemployment insurance or whatever, or unions and all that. I see the population of the poor increasing.

Historically, the unemployment figures in this county are always high. Some of it is due to the seasonal processing industry. The large work force of housewives during the summertime, according to the record, are all unemployed the rest of the year (which to me is ridiculous). Nevertheless, that's what the figures show.

The Spanish-speaking population has increased, and it's become permanent. At one time it was highly mobile. We used to have labor camps where they'd live to work during the harvest, then go back to their permanent homes. That kind of movement has essentially stopped. These people have settled down here and become part of the permanent labor force. Approximately 20 percent of the population is now Spanish-speaking. They are gradually being amalgamated into the general system, but the biggest percentage is still farm labor. More of them are going on in school, and with affirmative action efforts they're being integrated. Mexican-Americans are managing businesses and they're in the school system. They have opportunities if they want them.

Most of our farms in this county are family farms, regardless of size. There is some corporate farming, but not much. Some moved in during the 1960s but there is no increase now. The family farm is still essentially the backbone of agricultural production. Size is not really a factor, because a family can do what a corporation does in buying new land and increasing the size of its operation. They probably have more economic pressures on them to do that than a corporation, because a corporation has other resources. There are large insurance company holdings in the county — they bought raw land and developed it. Some Arab money has come into the county to buy land. And there are families that have got to the point where it's to their advantage to incorporate. But they're still a family farm.

A large percentage of grape-growing farms in this county are still under 20 acres. Many of these are part time. A 20- to 40-acre vineyard can be a viable economic unit and can maintain a family income. Actually, it's hard to pin down "viable size" to an individual crop, because there are very few farms with just one crop. They diversify for protection. But in most of this county, the "average" farm size (165 acres or thereabouts) would be a pretty good economic unit for a family to exist on without outside employment. The only reason the average is even that large is because we have larger farms on the west side. The 160-acre limitation wouldn't affect farming in Stanislaus County much — not here on the east side anyway.

From tree to lug box to tractor pickup, orchard fruit is still a labor-intensive crop. Reprinted with permission of the Department of Special Collections, University of California, Davis, Shields Library.

# A VIEW OF THE ORCHARD
## The Lind Family

On the road to the Linds' peach ranch are many small acreages of trees and grapes. Across from the modern metal farm buildings of this substantial operation is a brick house set in a pretty landscaped yard. Many gracious touches reflect the family that lives here — thick carpeting, wrought iron lawn furniture, and several healthy-looking long-haired cats strolling the premises.

Homer is a blue-eyed, graying blond man, gregarious and active. He likes to talk and has been elected to several positions of leadership in farm organizations. In a turtleneck pullover sweater and plaid pants, he does not look as if he now does much physical work. Louise, in a checked pants suit, seems businesslike and cautious, as a successful real estate broker doubtless should be.

HOMER LIND, AGE 62

My grandparents were farmers in Sweden and they came to Colorado to farm there. My father, who was born and raised in Colorado, left the farm and worked for the railroad in Oregon. My mother's father ran a commission service in the stockyards in Portland for many years. My mother and father were married in Oregon, and after I was born in 1917 moved here to California in 1920, where we've been farming ever since, on the same place. There were a large number of Swedish farmers even before 1920 in this area. My father came from a large family of a dozen children. All of them eventually migrated to farm near here.

At the turn of the century land around here was open wheat fields. With the establishment of the irrigation district, that land could be irrigated. We moved onto irrigated land and have been raising crops under irrigation the whole time we've been here. There have been changes, naturally. In 1920 we were very small, very diversified. We had cows and chickens and raised mostly crops like alfalfa for feed and beans and corn. It wasn't until the mid-1920s that we went into fruit — peaches and apricots — on a very small scale, a family type of operation.

We settled on 20 acres, and we still had 20 acres up to nearly World War II. During the Depression no one was buying land. You were holding on to what you had. We didn't expand until just before the war. When I came back after the war I added to the land my father already owned. In the late 1940s and all through the 1950s we got larger and went to all peaches. It wasn't until the 1960s that we started to diversify again. Then, keeping mainly peaches as our primary crop, we went into almonds, walnuts, and nectarines for fresh shipment, and some open farming in between rotation of tree crops.

The reason for diversification is different now. We have diversified to keep help over a longer period of time; the more work we can offer, the better help we're able to have. And with the cost of equipment now, we have to use it more days per year. We need to cut down our overhead by utilizing everything, both personnel and equipment, through a longer period. Before the Depression we were diversified because the family ran the farm. We didn't hire anyone especially, so we had to be diversified so that we could do many jobs over the year and not have any one job so big that the family couldn't handle it. My father and mother and sister and I did everything. We shared work with neighbors. Neighbors would get together at harvest time, whether it was hauling hay or picking peaches. Four or five neighbors — they'd all have 20 or 30 acres — would work together. People no longer do that.

Another reason to diversify today is that we're borrowing almost all the money we're using. If we have just one crop we're too vulnerable. Once, if we lost a peach crop, we lost everything. Now we diversify with five crops, hoping that at least we don't have a failure out of more than two of the five. Even that could be pretty harmful, with the amount of money that we borrow. Earlier no one really borrowed any money. In fact, I'm not sure where you would have gone in the 1920s and 1930s to get money. If you had a crop failure, you hadn't really put that much into it except your own labor.

When I was growing up my family supplied a lot of its own food. Everyone did, it was common. Today I don't know of one thing that we supply for our own diet except for the crops we raise commercially. The milkman comes to the door. And we have the same services and utilize the same conveniences as if we were living in the center of a big town.

It was typical of this area 50 years ago to have mostly 20- or 30-acre pieces. We had several big, big farmers — the big farmers then were <u>really</u> big. Today we have very few small farmers, unless they make their basic incomes from a job. People who make all their income from the ranch, for the most part, have to be fairly big. We don't have the real, real big farms anymore, but we do have a lot of medium to large farms. A medium-size peach grower nowadays would be 80 to 150 acres. It's almost like keeping track of inflation, it's hard to tell. Theoretically, though, that could earn a comfortable family living through full-time farming.

In spite of the fact that people had much smaller farms 50 years ago, in this area, I think, everyone lived fairly well. In fact, looking back, they probably in many ways lived even better than we do now. They were a lot more relaxed and had more time for enjoyment. We have all the conveniences, but we work so hard that we don't have the time to enjoy them. Neighbors don't socialize like they did then, either. Today you may not know your farm neighbors any better than city people know people next door. Land is changing hands so fast today that people are coming in, buying, investing money, who do not even intend to do actual farming, except for providing the money and hiring the people and having a manager take care of the operation.

I was old enough during the Depression to remember it, and yet I was still young. Like many others, I didn't have much money or many activities. I joined the California National Guard at age 20, in 1937. That gave me somewhere to go and also provided a little money. When World War II came along I applied for flight training and went through Randolph Field. I served all during World War II as an officer and a pilot. When I came out I stayed in the Reserve. I was recalled again in 1951 and served for three more years. I retired from the Air Force Reserves a year and a half ago, after 40 years.

My college training was not in agriculture, it was in aeronautics. Had there been opportunities in flying when I came out of the service, I would have gone to the airlines. But there were no jobs at that time. My only other practical experience was farming so I went back to it. My father retired from farming after World War II and became a real estate broker, so I just took over for him.

Our change from diversified farming to specializing in peaches was very gradual, probably over a period of ten years. None of agriculture at that time was as productive as it is today. On the other hand, we probably kept as much money then as we keep now!

I have both canning peaches and freestone peaches that can be used for canning, freezing, drying, or shipping. The history of the peach industry is ups and downs. We're on the top side now; we just came out of the low side. If you go back through the years, the ups and downs have come out about the same. The only difference is that since World War II we've been organized, with things like the Peach Association, the Advisory Board. Because of legislation we've had the opportunity to work together better.

One reason for cycles in peaches is because they are a tree crop. Even if the industry is in a surplus condition, it's hard for a grower to go out and push those trees out, because it's taken five years to start harvesting the fruit. You hate to take out trees that have reached their prime, just because the market's down. On the other hand, if you pull them out and the market comes back, it takes about five years to get your production back. It isn't like tomatoes, where you can plant 1,000 acres one year, and 10,000 the next — all you have to have is land that you can prepare and plant. We've never been able to solve our peach problems on the long term — what we've done has always been short term. We'll probably continue to have cycles. For instance, three years ago we had a surplus of about 10,000 acres of trees in California. We could use those 10,000 acres next year, but we can't carry trees for three years — a lot have been pulled out. We probably overreacted to the surplus. Now we'll probably be slow in reacting to the need again.

One problem was that in the late 1950s and the 1960s peach orchards were planted on speculation. Money was put up by the processors for growers to plant. Almost all that stuff is gone now. Processors aren't encouraging growers to go into peaches anymore. The growers themselves tried to regulate new plantings to some extent, but we never did get that legislation in California. We had the legislative tools for many years to surplus or pull out trees, but we never had the control to regulate new plantings. There is a real dilemma there. On the one hand, you can understand why people would say that you can't restrict plantings; on the other, if there's an uncontrolled number of people saying, "Well, I'm going to try peaches," that might just send the whole market five years hence into a spin.

The Advisory Board now is doing a better job than a few years ago. Instead of spending money for destroying fruit when we have a surplus, we're spending money to promote markets. When we have too many peaches for domestic use, we're promoting overseas. Those efforts seem costly, but in the long run it will pay off, provided we can go into markets without high tariffs. (Now we have competition with peaches from South Africa and Australia, and Greece is coming into the forefront pretty fast.) We do things by way of marketing regulations that will help sell more of our fruit in Japan, Germany, or England.

And we're trying to open up new markets such as Singapore, Korea, Thailand. China will probably not be a market for us. They're due for increasing income per capita in years to come, but peaches will be one of the last luxuries they'll be able to afford. Any country is going to buy necessities before canned peaches.

That's another of our problems. When there is a depressed climate for business, and money gets tight for the housewife, she will cut out things like peaches. A can of peaches is not a necessity, it's a high-price item, a dessert item. We're not providing a staple. We do better when our country prospers, or in countries that have a higher living standard.

Crops in peaches are quite variable from year to year. In early spring, frost can take your crop anywhere from the time the trees bloom to when the pit hardens. A big problem is rain at harvest, when brown rot starts. Some winters we don't get enough chilling, and the buds don't set well. A couple of years ago we had a shortage of water. And so on. Any number of things can spoil your crop.

Some growers in this area who were pioneers in peaches in the 1920s don't grow peaches now, though they are still farming. They've chosen to go to a nonperishable crop, or a crop that adapts itself to mechanization. There has been a real shift around our area. Some ranches here in the 1920s and 1930s were all dairy, and then somewhere in the 1940s and 1950s they were all peaches. Now, since sometime in the mid-1960s, some have gone into almonds or walnuts. This particular area is real good for peaches — the best in California. But it's also real good for almonds.

And then we're losing acres every year out of the county to asphalt and cement. A map of this road today would be radically different from 1965. When I say farming acreage is lost, I'm including "ranchettes," three to five acres. Those might as well be asphalt and cement as far as producing a crop is concerned.

Another of our big problems with peaches is that it is a high labor crop. We have a new element in labor — the only way you can say it is "Chavez." We're dealing with labor unions now. We can be the primary party, or we can be the third party, like two or three years ago when the canners were negotiating with the cannery workers during peach harvest. We are the raw fruit producers, and we had nothing to do with those labor negotiations, and so we just sat there while they took their time. Well, processors don't pay for any fruit until it's received. In our contract it says that we will not deliver again until a strike is settled. So we have no way of pressuring either side or helping either side. We have to stay neutral — and yet we're in a period of time where we are extremely vulnerable. Anything that becomes overripe, we have to let it drop on the ground. And we absorb that loss, the cannery doesn't. Years ago we never had this type of labor interruption. In a perishable crop you can have a disaster. If you have varieties of peaches that come ripe in a short span, and the strike lasts long enough, you could have your whole year's production wiped out as a third party, while the two primary parties try to negotiate a price.

Anything that needs labor to bring it to production has a high cost. Contrast peaches to almonds: If you don't harvest almonds this week, you can harvest them next week. The grower himself will probably be going out to turn that key on and off, whenever he wants to start up. Harvesting almonds is all done with machines; you have complete control. The only problem you have in starting harvest is whether the machine will run. It's not whether the workers have arrived, or whether the workers are happy, and all the other things. The management problems and other conditions you have to meet, such as OSHA and pesticide requirements, complicate life when you have many people in your orchard, versus one person on a machine.

In our operation now, our labor needs are almost throughout the year. We need workers in the winter for pruning and cultural practices, mainly spraying. We need workers in the spring for cultivating, irrigating. If we've got too many fruits on a tree, we've got to take some of them off and space them. Thinning is all done by hand. Then we move into harvest. Clings are now about 70 percent hand-picked and 30 percent machine. The experts figure in about ten years it will be 30 percent hand and 70 percent machine. They are trying to develop varieties more adapted to machine harvest.

I have about 150 acres of peaches. In October I'll have just two or three workers. Going into winter I'll have from 10 to 15. In May we start getting busy with the thinning, and I'll go up to maybe 30 or 40. At peak harvest it can be up to 60 workers. This is for peaches only. But between thinning and harvest of peaches, I also have nectarines to harvest and ship fresh.

Fifteen years ago I had almost all white help, local people. I'm down to only two local men, who will stay with me until they retire. The younger white people now are in processing plants driving forklifts, in machine plants with wrenches as mechanics. Some are building houses. Some have gone to college and are working for the government or in other professions. The only white help we have today are those that were too old to make the transition to anything else. As they retire, they're all being replaced by Mexicans. Most of my workers come from one or two villages in Mexico. Some have been with me 10 or 15 years. I can name almost half of the people who will be here in May. I'll never receive a letter from them, and I won't write to them. But they'll drive into the yard, and I can almost pick the day that they will come. Some additional people come from Mexico during the summer, because they have kids they don't want to take out of school. Most of my employees work only for me, but in some cases, if they want to work longer after they finish here, they'll go into grapes until they are ready to go back. It's a pretty stable situation.

My definition of the family farmer is that he is the man who still has personal contact with most of his workers. They know exactly who's writing the checks, versus the corporation or large farmer who may never see his employees, who may never leave his office other than to drive out alongside the road — who has no personal contact. He probably gets very few worker returnees unless he has a good manager. Even then, he'll have a higher turnover and probably will hire more workers who are not legal. But I can't afford to hire illegals, for the most part, because I can't afford a raid. I'm not looking for cheap help anyway, I'm looking for good help. In my case, most of my workers know not only that they are going to come back to work for me, but also where they are going to live, because they make these arrangements every year. They live in privately owned places, or they use seasonal government housing in three or four places in the county.

I'm one of the old-fashioned farmers — I have never mechanical-harvested one peach. I am not looking toward it in the future, even though the time will come when I'll probably have to. I have not ordered a machine, and I'm not training my trees for mechanical harvesting. With good help I can offset my higher costs by better production, and I choose to go that way at least for now. But I have neighbors right across from me who are the opposite. To say that they're right or I'm right is the $64 question.

If mechanization does proceed generally in peach harvest, we'll have a most serious problem. My workers go back and forth, but as time goes on a certain percent stop going back and forth. As I get a little larger, I can give some of them work for maybe almost a year long. But if more stay I won't be able to keep them. Then if I do start cutting back, or gradually go mechanical, these people would not have anywhere to go, because my neighbors are cutting back too. Those who have decided to stay here may have to go on welfare. Urban people won't understand this probably until maybe five years after it happens. Then they'll wonder why the welfare rolls are so high. We can't train all these people for jobs, or move them all from the country to the city, or from the farm to industry. Industries are also becoming mechanized. As we need fewer people in our work force in California, many of these people will be untrained. If the government trains them, they may still not have a job. In California especially we've had more farm workers come here in the first place, because we have the specialty crops and also pay good wages. They're not about to leave if they've established themselves, and yet they may not have a job in the future.

When I started school in the mid-1920s, the population around here was all white, predominantly Portuguese and Italian. That probably held true up to the 1960s. Our public school here has now taken a turn, and I would say that we're approaching 50 percent Spanish-speaking in the grade school and one-third in the high school. The percentage of Mexican-Americans is moving up fast, because they have big families. Many of them want an education, and many are real good students; but there is not enough year-round employment for all the families. That's the reason this county is high on welfare, high in unemployment.

A few years ago the United Farm Workers (UFW) tried some organizing efforts in this county, but only on the west side. On the east side we've been pretty well left alone. We're too diversified, we're too small. Even if the workers were organized, they don't have the opportunity here for unions like they have in the Imperial Valley, where 300 or 400 people may be hired at one ranch out of one camp and might work for a good part of the year. The union makes its efforts where more people are needed. It's a more convenient target for union organizing.

My wife works off the ranch. A few years her salary in real estate was greater than my income on the ranch. I can tell you what's making things brighter right now. It isn't what we're making on the raw products we raise, so much as the fact that we can borrow more money on the land we have. When you have had land a good many years and are paid up on it, you can borrow a considerable amount of money. Land inflation has done more to bring farming into the black than anything else. If a person three years ago bought open land for $2,000 or trees for $3,000 an acre, today the open land probably would be going for $4,000 and the trees would be $6,000. Instead of selling, he could go in to the bank and borrow the extra $2,000 or $3,000 added value to the land and use that for almost anything in expanding his farming operation. If you've got the land, you've got the equity to borrow with.

That just about means that a young man without money from parents or another family source — unless it's money from the Far East or somewhere — can't get into farming. To start with, if he comes in at $6,000 an acre for peach land, and I'm farming next door to him and I've had mine for 30 years, if you just figure the interest he's paying on this $6,000 on today's market, versus the fact that I'm paying no interest, right there he's under a handicap.

People who worked for my dad a few years before the war are now farmers in their own right and are retiring. Many of these people came from the Dust Bowl and worked 5 to 15 years and were able to save up enough money to buy their own land. They did it with hard work, which they could do then. But today it takes more than work; it takes an incredible amount of money. That's a problem for the Mexican farm workers now, because they're never going to have that opportunity. Or at least it will only be a very lucky few.

The great upward pressure in land prices in this area is due partly to urban growth, but it's also competition from insurance companies and money from professional people, mainly doctors or lawyers or airline pilots. Many high-salaried people in the cities want to find some place to put their money. They come in with tax money, it's just that simple, and raise the price of land. They aren't buying the land to make a living on it, they're buying the land to avoid taxes. They're also hoping that in a short period of time the land will inflate and they will get additional money back that way.

Probably our number-one problem in the family right now is how to transfer our property to the younger generation. We really don't have an answer. We're looking into it. My mother just passed away, and my father passed away four years ago. Most of that property transfer was done ahead of time, and at different land price levels. With me being the only son, there really wasn't much problem. But with our four boys, looking at tomorrow, we're looking for answers to the problem of transferring an estate to them. We have an attorney, and I've talked to a lot of people. The family corporation is, for some of those who have looked into it, the best way to keep an estate together. We may go that way. The son who farms with me now is slowly taking over in his name a part of the land. We're farming in cooperation with each other, but it's not a partnership. My second son is leasing land this year for the first time. I am in partnership with my other son in avocados in another county, which we just began last year.

Family farms will survive one way or another. Many of them could sell out today and be well off with the inflated prices. But if we end up with corporate farms, food prices in the nation will go up. You can only get so big in farming and then you lose your efficiency. Actually, it's individual how big a farm should be. I know people who are efficient on a fairly large scale, and people who are inefficient on a very small scale. It's more a matter of management than size.

I've always been active in our local Farm Bureau. At one time we had "farm centers" in every community. Most of the activity in the small rural community rotated around the farm center and the various churches. After the war it started declining. Farm people became active in other organizations like the Lions and other civic associations, and they joined commodity groups like the Peach Association, the Tomato Association. There were more things to do. People moved in who weren't quite as community-oriented. We finally had to give up the farm centers in the county somewhere in the early 1960s. Our county Farm Bureau board, which previously was made up of community center presidents, now is made up of directors who are elected by the county membership. The chief functions of the Farm Bureau have changed too. We are more political, we have more lobbyists in Sacramento. We spend more time with our legislators, invite them to our functions. Like all other industries, we have to be more in touch with the people making the laws that we have to live by.

The Farm Bureau has a large membership nationwide. The state organizations vary. California is completely different from, say, Kansas. You have one or two commodities in Kansas, but in California we have many different commodities. A peach grower doesn't necessarily have the same problems or interests as a tomato or a walnut grower. But even though we are different, we have to come together. Pesticide regulations are common to all of us. Labor is common to all of us. Land use is common to all of us. That's why the Farm Bureau is becoming more political. We used to have most activity in the local community, now we have much more activity at the county and state levels.

Urban people don't realize that agriculture does more than feed them at the cheapest rate anywhere in the world. The American public has been spoiled. They have a second car and a second TV and are able to spend weekends in their boats because they've been able to buy food so cheap. As food prices go up, there will be less buying power for the other nice things. Agriculture has a greater bearing on the economy, on the future of the country, than anything else.

## LOUISE DAY LIND, AGE 56

I was born in 1922 and grew up six miles from where we live now. My family lived on a small dairy farm on the outskirts of town. I came from a large family of seven children. We all went through school in Kinsey. Eventually I went to junior college. Some of my family went on to the University of California. Now they're scattered pretty much all over the world. I always said I wasn't going to marry a farmer (when I was 12, anyway) because I'd lived on a farm all my life and I thought it would be glamorous to live in the city. But I did marry a farmer, and so I have spent my whole life on the farm except for the few years that we were in the military.

Our dairy was just a small family operation. We also had a few nut trees. My father sold some produce at the local markets. The dairy was the one thing for regular cash flow. We had that for many years until my father passed away. None of the children took it over. One of my brothers did go into dairy farming elsewhere but decided to give it up and went into almonds, and has been in that ever since. He had a Grade B dairy, and did all the work himself, which was very hard. It never got to the point where he could hire a full-time man. It was a matter of getting bigger or going out. He decided to go out.

I went to college for a brief time at the beginning of the war. Those of us who had a good high school course in stenography were asked to take tests to go into civil service and work with the military. So I worked as a secretary at the air base nearby for two years before I was married in 1943. Homer and I had known each other quite a while. We lived in Tucson, Arizona, for two years during the war before we came back to farm.

I was never really involved in the farming operation myself. I did small chores all around, but I never have gone out to the field and worked. I kept the books and that sort of thing, and raised my children. We have always been active in the church, and we were in PTA and followed school activities closely. We never missed a football or a basketball game.

My father-in-law was in real estate for about 20 years after his retirement from active farming. Some years ago he encouraged me to come in with him. He kept encouraging me till I became a broker. I've been successful at it, but it's a matter now whether I really want to continue the hassle of being in a business like realty. Things have taken a turnaround. I prefer to sell farm property, but I sell anything I can — we do sell some homes. It's just a small business in a small town. I think about retiring every now and then, but with my last child off to college and the house kind of quiet now, it's rather a foolish time to quit. It's harder to get good things to sell, though — prices are so high. Up until recently, real estate in this area was turning over rapidly, but the last 18 months have been much slower. Many more people are in the business, and that makes it difficult too.

There's quite a demand for small acreages. We can't possibly find places for all the people who want to buy "ranchettes," or 20 or 40 acres of almonds, or whatever. When pieces do come on the market, if they're anywhere near market value, buyers in the immediate neighborhood will buy them without a broker. Farmers tend to keep expanding. Also, about five years ago some county people decided that they must save farmland for agriculture. They put in new zoning laws prohibiting small acreage splits. They rezoned this area into A2-10 parcels, which means that we cannot split parcels into less than ten acres. That stopped the subdivision of small acreages, but on the other hand it made those ten-acre parcels and previous splits go higher and higher, because they're scarce. It has made land values go up, because it's put so much pressure on what's already there. If it hadn't happened, though, we'd have been swallowed up by small parceling for country houses.

In a way, I'm surprised that our sons want to stay in farming. We worked really hard, and we didn't thing we were raising any of them to be farmers. The one who has come back was away ten years before he returned. The other two are in farm-related business. The youngest one is studying to be a C.P.A., but he's also selling real estate and has a license already, even though he's only 20. I think the one thing we presumed our children would do from the beginning was go to college. That was the uppermost thing in my mind. My teachers had wanted me to go to the university in the worst way, but there was never any money for me to go. So I was determined that my children would. When they were growing up we didn't think that we could possibly send all four to college, because there was never that kind of money around, but none of them has ever missed a semester at college. Education is really important for them, because farming is not just growing things anymore. It's big business, it's complex. They need all the education they can get to survive in farming now.

Carl and Anita are a striking young couple. He is tall, slender, his blond hair curling out from under the edge of his farm cap. He looks like an ad for a breakfast cereal, the all-American boy. She is small, dark-eyed, with gleaming long black hair, faintly exotic even in blue jeans. He speaks softly. She expresses herself with more determination and is frank about the adjustments she has had to make as a city girl moving to the farm. But when the baby wakes up, she melts into a fond mother, cooing at the tiny girl who is a miniature version of herself.

## CARL LIND, AGE 33

I was born in 1945 in Tucson, Arizona, when my father was stationed there during World War II. We moved back to California within a couple of months after my birth, and up through high school I grew up right here. I spent two years at the junior college, three years at California Polytechnic, and three years in the Army Corps of Engineers. I was stationed in Thailand, Germany, and Viet Nam. After that I went back to Cal Poly and received my Master's degree in Business Administration.

From there I took a job with Bechtel, the big construction company, and spent about eight months in Los Angeles before I decided that the farm was the best place for me. So I have an engineering degree, but I couldn't say that it would be a livable profession for me now.

In high school I had an ag project and took ag courses, but I always thought farming just wasn't where the money was. The hours were too long and there were too many risks, and I thought there was a better life out there somewhere else. Engineering seemed to be a growing field. Without really knowing too much about it, I spent all my five years of college becoming an engineer.

My father had always encouraged me and my brothers to go out and discover new things, and even during the summers to get jobs off the ranch. One summer I worked for an agricultural chemical company, which gave me a broader view of agriculture because I was exposed to different types of farming for the first time. I worked for the Cling Peach Advisory Board one summer, and for the Fruit Exchange one summer. A couple of summers I was in ROTC. My dad wanted us exposed to everything. I sometimes felt that he would prefer that we didn't choose agriculture. He himself was a pilot, and his love of flying was great. I know that sometimes he wished that he were flying. He wanted us to explore different avenues and possibly find something we liked other than farming. You see, the 1960s were very tight years in the peach business. There was more supply than there was demand, and prices weren't good. Farmers all over California were forced to dispose of part of their crop through a green drop program in their marketing order. Along with the usual weather problems, it was a very discouraging time for farmers.

When I was working at Bechtel I just got tired of sitting behind a desk. I made the decision to quit and come back to the farm. In the Army I had been outside all the time working on projects. At Bechtel it was not the same. Bechtel was an impersonal, very big company, one of the biggest privately owned companies in the world. You can become very lost in an organization like that. Farming, on the other hand, is very personal. And there are few other professions where you can be outside and enjoy it as much as in farming. I made my decision four years ago and I haven't been sorry.

The outlook economically was better in farming then than it had been for a while. Part of that was the weather, which had turned crop prices around. But there was still a great deal of uncertainty in the cling peach business. It hasn't been until this last year that things are really changed. There is no way now with the number of trees we have in the ground that supply can exceed demand. We can go into the next three or four years knowing that there will be a home for our product and a chance of profitability. Like anything else, though, if the price of peaches is too good, growers will be replanting, and the supply again will go out of balance.

I rent part of Dad's ranch from him, and I hope some day to buy that portion. Right now I'm growing peaches, just a few walnuts. I haven't spread my risks yet. Dad has rented me one of his better pieces of ground, with mature trees. He's really given me a good hand there. If the next few years show a profit, I may be able to start buying the land from him. It would have to be from him, because the price of agricultural land in this area has climbed completely out of sight.

I had more free time before I began farming. When I worked for Bechtel and even in the Army, at least at night my time was my own. Here on the farm, especially during late spring and early summer, it gets to be almost a 24-hour operation. It's hard to get a weekend off, hard to give the appropriate amount of time to your family. There's work to be done from May on, when the canals start filling up and irrigation starts. Irrigation becomes very critical and timeliness is important, and it's difficult getting good labor. When we have extra work during this period, we do some of it ourselves. Irrigating is hard to train somebody to do. We have some very good men working for us, but when we get into thinning and picking, their time is taken up running the crews. The odds and ends we pick up. It has to be that way. Some things you just feel more comfortable doing yourself. I'm not a desk farmer anyway — I abhor paperwork.

When I was growing up around here you could leave your door unlocked and feel fairly secure that nobody would come in. That's changed. Since I came back to the country five years ago I've been burglarized twice. On three different occasions people have taken things out of the yard or the garage. The crimes in the city are happening now in the country, and with a great deal of regularity. Much of it is juvenile crime. I used to think Kinsey would be one of the last places to be caught in the problems that city schools have — drugs and violence and racial tensions. But the mix of the town is shifting even here. There are more minorities and problems with adjustment.

The labor force on the ranch varies from month to month. The peak is probably about 80. It's very cumbersome, having to work with that many people. To be honest, we would rather not be relying so heavily on labor. I feel very vulnerable, especially during harvest. We haven't gone to mechanical harvesting, like some farmers. We have nectarines, and we need a certain size crew since they can only be picked by hand. It's not easy to get a crop to mature 100 percent just at the right time so you can go in with machines. We normally send our crews in and pick off 50 percent and let the other 50 percent mature or size up for a later pick. Mechanical harvesting is just a one-shot deal, regardless of what they try to tell you.

Workers do jump from one place to another sometimes. Because it is piecework, sometimes they'll hear of a friend making more money on another ranch and say, "That sounds pretty good, I'll jump over there." The smart ones stay with one ranch and go through the good times and the bad. They make more money because they get more weeks of employment. They may not make quite as much during one week as their friend over on another ranch, but if the friend is done picking and has no place to go afterward, then he's not making anything.

When the government started getting into regulating workers' safety, we thought we might have to go to an all-male force. Women weren't supposed to climb ladders past a certain point, or something. We could have lost many workers that way, because a number are man and wife teams. The regulation changed, though. Generally, the average age of the field worker is between 30 and 40. Most of them are family men and their wives and they have young kids. We couldn't get our crop harvested without them. But we don't allow children under 18 in the field.

Local teenagers used to work in the orchards, but I don't think you could get a teenager today to go up a ladder. It's pretty hard work and they seem to have an ample supply of money elsewhere. We used to use teenagers as "swampers" too. They loaded up the lug boxes from the pickers onto a palletized trailer, then the trailer was taken into town. The trailers were unloaded by hand too. It was quite a chore, but it was kind of fun to come out, and they got a good day's work. Then the forklift came along. Now we don't have "swampers" at all anymore — we went to bins, which take the place of 24 lug boxes. It's just so much easier. That was one of the first real breakthroughs in mechanization on the farm that I can remember. We used to hire three times as many hourly workers as now, because it took a lot of work and time to get those boxes loaded.

There's been a shift by consumers toward more fresh fruit and less canned, I suppose because of the health food movement. People are conscious of the amount of sugar they're taking in. In the Bakersfield area enormous amounts of land are being planted to fresh fruit. It's going to be hard for us here to compete with that because they get an earlier jump on the market.

My dad is 62 now, and he shows no signs of retiring. He still makes the final decisions. We all put our ideas in, but he shows no signs of letting go the reins. My mother has made a tremendous contribution to the overall ability of my father to farm. I don't think she realizes sometimes what she has contributed to this family operation. She's always been the stabilizing force in the family. My dad's had a difficult time raising four boys, because we're just as opinionated as he is. After we were pretty well grown, my mother went into real estate. During some of those years when it was tough to make a buck, she was the one who actually made the money. She's not an aggressive real estate person, but she has a reputation of being honest and fair. Some of her sales came during the time when peach prices weren't good. There were a couple of years the ranch didn't make anything, so her money carried us through.

Friends of mine are not farming today because they haven't been able to establish satisfactory relationships with their fathers. The relationship I have with my own father is rocky at times. I'm more willing to try something new, experiment with new cultural practices, try a new product or make a change. He's been resistant sometimes. Sometimes I've proven that I was right, and he's followed up and made a change on his part of the ranch too. This is common between generations, but it makes it difficult for some young farmers to get into the family operation. Some strong-willed older farmers out there are just difficult to work with. One of my friends, who served time in the Air Force and saw some of the outside world, tried to go back to the farm and found it impossible. His father wasn't willing to bend, and they couldn't negotiate the adjustment.

More than anything, farming is management. I'm personally not mechanically inclined — my wife will say that! Some farmers work on their own machinery, but I don't. Today's farmer has to be more of a manager than a mechanic. Even though I hate bookwork, and I don't like to keep papers, I know that being a manager is the most important thing.

Farming is going to be very hard in the future. We're going into bigger business all the time. Somebody said just the other day that within a certain number of years we'll probably have two large corporations in the United States running everything, including farming. That's not right now, and maybe not in the next generation. But it could be two generations away.

## ANITA SALDANA LIND, AGE 27

I was born in Los Angeles in 1951, the oldest of six children. My father has French and Mexican ancestry, my mom's Mexican. So I'm Mexican-American. My dad was born in Los Angeles and has always lived there. He's a foreman with a paper products company. Actually he's more interested in real estate; he owns over 100 duplexes down there. My mom is a typical mother. She always stayed home with the kids and helped my dad and was his general support. She helps manage the duplexes — she collects the rent, she finds out if they need windows.

My parents always stressed education, so it was natural that I would go to college. I went to junior college and then to Fullerton State where I took speech therapy classes with an emphasis on psychology. I took business, too, because my father expected me to go into real estate. But after four years of school I wanted to see what else was going on, and so I stopped to work for a year. I worked as an engineering aide, something completely different from what I'd ever done. Carl and I met in October 1973 and we were married in June 1974. It was fast. And we came right up here.

When we were engaged I just assumed that we would live down south and Carl would be an engineer. But afterward he decided he wanted to come back to the farm. We came back here immediately after our honeymoon. That was really the first exposure I had to farming. I had come up once in May just to meet his family, but of course I really wasn't looking at it objectively — I was just with my fiance. It was all sort of overwhelming.

Farming is nothing like I expected. People sometimes have the idea that a rancher's life is just free and easy — grow your garden, and do your own thing, and have an out-in-the-country type of life. Well, I had no idea. Especially, you know, peaches. I never realized that farming was that much work! Besides that, it's a business. I never realized how much you had to have, even to get into it. The amount of time that Carl puts into it awes me, and the information he has to have for dealing with business matters, and the types of people he works with to get this or that. He has to deal with payrolls, he has to deal with workers. He has a relationship with people that is just totally different from merely knowing materials. There are so many elements that are important that he has no control over — like nature itself, and the canneries, and the work force that isn't always consistent. I still was geared to a 9-to-5 day as far as working was concerned. Sometimes I see him now waltzing along with only three hours of sleep for five or six days straight. It really goes like that sometimes, oh yes.

Because of irrigating and everything on top of that, they put in many, many hours. I wasn't prepared for that. I had my own idea of what my husband was going to be like, and what my home life was going to be. All I had to judge by was what I knew at home, and the situation we were in before we got married. So it's been quite an adjustment for me. I have missed urban life, too. I don't like this valley at all. I don't like the weather. The summers are too hot, the winters are too cold. We both have hayfever, so spring is just miserable, and in the fall when all the Bermuda grass is drying up, it's miserable again. People say, well, you're so close to San Francisco, and you're close to the coast, but two hours' drive is really two hours' drive. It's four hours when you're commuting back and forth. We don't just hop to the Bay Area for a dinner and a show.

I like some things that just aren't offered here. I love plays, for example, and the plays here (they do have some) are not what I grew up expecting. We lived right by the Forum in Los Angeles. The college has been a resource, but then again I came from a bigger college where there was much more offered.

But now that I'm here, I've learned to live with things. Some things I do find enjoyable. Ever since I moved here, I've gone to school, and that has taken up many hours. I got my B.A. here and my credential. I got my Master's. I've been on a schedule where I work all fall and all spring and then have the summer off. Right now I'm working, not because it's a monetary thing (although it became that since we bought this house), but because I want to be accredited from the state, and you have to work for nine months before accreditation. I work for the county schools as a speech therapist. I travel back and forth to different schools assisting children with their speech problems. I don't intend to continue that, though. I'm very traditional in this sense — I believe that if you have children, you should devote your time to them. I've already missed some of my little girl growing up. She was born last spring, so she's less than a year old. I was working up until the day that I had her, and then went back to work in the fall. In a couple of months I will be telling the school district that I want part-time work, but they've cut out most part-time employees, so I'll probably be staying home. I'd like to have a private part-time practice. There's a need for it. Then I could have my own hours and my own days. It would be like a tutorial service.

On the farm I've done the payroll and some bookkeeping, but at irregular intervals. I never really knew what I was doing, never really saw the end product or what the whole operation was. We've tried a bookkeeping service, and that's probably the best way to do it. We have paychecks for 60 people, and they've all got different names for various reasons. You've got to keep track of numbers and names, and each one makes a different rate of pay. It gets really complicated.

I don't really see any active part for me in the farming operation. I might go out and irrigate with Carl once in a while, just to keep him company — I've done that in the past — but I would not take on any responsibility. I don't particularly enjoy being out there, so I wouldn't want to put in the time.

Carl had a set of friends when we moved here. However, resenting the fact that I was taken away from an environment that I liked, I didn't encourage much social activity. But I've grown out of that. Now we have friends that we enjoy playing racquetball with, and we do things with a cousin of his. But we don't socialize much, because there are great periods when we don't have time for recreation. Our schedules have not been very compatible. We enjoy being together on weekends mostly. He's fairly involved with his church, but he's Methodist and I'm Catholic.

I have felt like an outsider, sometimes, coming into a family business. I don't think it's anybody's fault but my own, though. I'm simply not interested enough to learn agricultural terminology and know what's going on all the time. Sometimes I feel like I'm not participating. I'm not as traditional as his family would like me to be. When his grandmother was alive, she would often say things like, "We'd like to see you in church more often." The family used to have quite a few dinners, and sometimes Carl would go by himself, because I didn't want to.

I have gone back and forth to Los Angeles to visit my family. That helped me. There are phone calls twice a week. A brother is up near here now, and a sister flies up every other month. She likes the area. I don't feel cut off from my family at all.

We just bought this house a month ago. We thought we'd better invest our money and start getting settled down. This place is ten acres, nine in young walnuts. This is the first time we've had land of our own. I do think a farm will be a good place to raise a family. I love the Los Angeles area, but when I go back there I see that the kids are exposed to much more than I was 20 years ago. I wouldn't want my kids to go to a large school like I did, with the competition and the problems. It's a more wholesome environment here.

Today's automated henhouse is a highly professionalized, cost-conscious operation.

# THE MINISTER'S EGGS
## The Lowe Family

What used to be a narrow country road is now a street leading into the city of Modesto. Small acreages dotted with 60-year-old barns are interspersed with expensive new brick country residences. Modest orchards and vineyards surround modest older homes, but signs of change are clear: Less than a mile from the Lowes' chicken ranch are the traffic lights of a large intersection with a shopping center on one side and a subdivision on the other, still under construction.

At 75, Otis Lowe is stocky, healthy-looking, with a full white beard. Occasionally hesitant in speech, his forehead wrinkles as he recalls the hard times, but his home today is a large, comfortable, brick-faced modern ranch style. Elizabeth comes in from pulling weeds in the garden. At 71, her face is still smooth and pink, her graying hair tucked neatly beneath a prayer cap. Her serenity makes her a beautiful woman. She knits on a bulky brown sweater while Otis talks.

OTIS LOWE, AGE 75

I was born in 1903, on Jordan Road north of town. That place is still there. My grandfather came from Virginia. He lived two or three years in Kansas, then came to Coalinga in California, took up a homestead and lived there nine years. In 1903 my parents came to this area. The irrigation district was started, that's the reason they bought here. My father farmed there as long as he farmed. And that's all I've ever done is farmed around this area.

My father had 40 acres, mostly alfalfa and dairy. He raised some peaches. He used to sell to Pratlow Cannery, I remember. Of course, then you had a few pigs and a few chickens. They said, "Don't put your eggs all in one basket," you know. I guess it was a pretty good idea then. When it come to irrigating, my father was the first one to strip check. They used to make square checks and make them level, but my father just strip checked. He'd let the water flow. It was fine for alfalfa and crops like that, but for some types of land it wasn't the best either. It didn't hold the water long enough for it to penetrate far enough.

They bought where they thought the ground was the best, but it was where the alkali come up first. It really just about ruined that country until they lowered the water level. Now it's good farming again because they've washed that alkali back down.

I was eight years old when my family moved to Drewsey for four years. Up there we raised oat hay. My father went there to get away from the irrigation. Irrigation was too laborious, and he decided he liked the way they did it in Virginia, I guess, and he thought he could do that up there. But we came back to Shelton and lived there four years, then my father came back to Ten Mile Road and had the 40 acres there until he retired. Them days it wasn't like in the East where they lived and died on the same property.

We thought we was well off, if we had plenty to eat and enough to wear. As far as making any money, we never got rich, but we always lived, we was always happy. In my family was four boys and four girls. We always worked together. My sisters didn't work much on the farm. They worked out when they got old enough. My youngest brother died at 23. The other two, they farmed, but their main income was always working out. They each used to have 20 acres. Used to be able to make a living on 20 acres, but times got to where you couldn't. Today you can't.

I was the oldest boy. I helped my father the most. We had two sets of teams. I generally had one and he had the other. I just had grammar school, didn't go to high school. Started working full time like an adult as soon as I was out of school. I stayed home till I was pretty near 23.

After I was married in 1926 we lived in different places. We didn't live over a year or two at a place, though. Renting, you know, that's the way it was. I never owned a farm until 1940. I never did go too much for working out steady, though I did work out a lot. But I generally tried to keep farming all the time, even during the Depression.

There were lots of places to rent. Them days a place might be sitting idle, nobody farming it, a house setting there, doors not locked. Nobody'd done any damage to it. They didn't think about locking anything. Somebody was living someplace else, making a living. Farming was just not good enough, though taxes weren't very high. Bank of America owned a lot of farms when they were lost during the Depression. You could easily rent a place from them. I never did, for the simple reason that people always warned me against it. The bank'd make you a good proposition, what they told you, and you'd get in and improve the place, and then they'd get a buyer. If you had it looking nice, which you wanted, to make the crops look good, well, somebody would come along and offer the bank what they wanted for it. The bank would say, "Okay, you have first choice," but you wouldn't have the money. So they'd sell it out from under you, which you can't blame them — but it sort of took your heart out of it. I kind of avoided renting from them for that reason. Generally the ones the bank foreclosed on were broke, though, and glad to get out of it.

My father died about 1946, six weeks to the day he would have been 80. He quit farming pretty quick after he was 60 and sold his land. I started farming in the 1920s, borrowed money and bought cows and thought I was doing all right and then the Depression hit. In 1929, that's about the time that my father quit too. Things got bad. Nothing was worth anything. I lost everything I had. That's the reason it took me so long before I ever bought a place. I couldn't get out from under the problems I had.

We had 42 head of cows in the dairy. We had a milking machine. We rented the property. My son wasn't old enough then to help. I went into it with my brother, who was younger than I was. I had about 19 head clear, then together we mortgaged them and got 42 head. We had them about half paid for when the Depression hit, and then they weren't worth any more than what we owed on them. Then we just quit. My brother went to work wherever he could, but I always kept a few heifers. We didn't pay bills. If we got a dollar, we'd eat it. Most of the big companies came out and said, "We've cancelled your bill, but just pay from now on." That helped. My uncle had tomatoes to sell and he told the boys, "If somebody comes along with a dime and says he wants more tomatoes, give them to them. But," he said, "get that dime." That's what money was worth then, not like today. We were able to provide for most of our family's needs by what we grew on the farm. We always had a garden. We never went hungry. We didn't always have what we thought was the best, but we lived, we all worked together.

Dorothy and I had four children before the Depression was over. My oldest boy George worked at home more than the others. The others worked out pretty much. Always said I taught them to work, and then the other fellows wanted it.

People couldn't get boys that would work. You know how that is. All my boys knew how to work.

Small farmers didn't go under during the Depression, the ones that borrowed money and operated big, they're the ones that went under. The little fellow on 10 or 20 or 40 or 60 acres, if it was paid for, it didn't seem to bother him. If he was out of debt on the land, why, he just took less income. But I hadn't bought a place, and I lost all I had. I lost the cows and wound up in the hole, and it took me ten years to get out. I rented a little land and worked it at home, and I worked for neighbors and all. But I didn't get a steady job. Them days there was a lot more men than there was work. I made a lot of crates, melon crates — if I had a profession, I guess that would be it — I'd be a crate maker. That is a lost art today. They make it with machinery nowadays and use cardboard and other stuff.

Times started to get better for us about 1940. I rented a place up in Sauerville, tried to dairy up there, but it seemed like money wasn't worth anything. I finally found out that if I raised sweet corn for the market, just for local stores around, I could make more. I was limited, of course, and couldn't go in big at that, but I could make $150 an acre at sweet corn, and that's when we started doing a little better. I supplied the stores in Modesto with sweet corn for five or six years. My children were still small yet, so my wife helped. I'd irrigate at night and pick corn in the daytime. She'd help gather it, sort it. We'd throw it in piles and our boys (Michael was about six and George was eight) would take gunny sacks and drag it out. After we got a little bigger, we'd just take certain aisles with a wheelbarrow. At the last we backed in there with a pickup. Our sweet corn lasted all summer. We planted every week so we had a continuous crop. We sold to Safeway, the main chain store then, and to the other local stores. Did the delivering and selling myself, and the picking and everything. I had a good market. Where I made a mistake was, when talk began to seem like times were getting better, I thought I'd quit. That's about the time the price went up. I sold a lot of good corn for 18 cents a dozen. I had a few other things, tomatoes, but them days I couldn't get even 50 cents a lug for a nice 30-pound lug of tomatoes. There were just so many others, and money was so scarce.

In 1940 I bought 20 acres and went into the dairy business. It was a mistake in the long run. We bought with borrowed money from a friend, $1,000, and put half of that down on the place and the rest I used for improvements, because you have to have a little to work on. We lived there from April to July, and then the house and everything we had burnt. We had $1,150 insurance and we put all that back into materials, and people helped us build a new house. We just finished the shell, really, we didn't have any doors or cupboard doors. In the upstairs we just had flooring enough across for the children to sleep on. It took us a good many years to get the house finished. The fire was really a blow to us. But at least then we had a place of our own.

Got rid of the cows in 1945. After that we got dairy goats. Our second boy bought some purebred goats from a neighbor, and then another man with 200 goats wanted someone to milk them for a couple of years, and we found ourselves in the dairy goat business. We thought it was a coming industry. And it is yet, maybe — but it got to where the market filled up. They condensed it for medical purposes, for allergic babies who couldn't eat cow's milk. But when the buyers got enough of it they would just say that they couldn't take anymore. Well, you can't pour it down the drain, but that's all you could do with it. When they finally cut us down to 50 head, why we got out of it.

A condenser named Rostock had encouraged us to go into it. That's where we went to sell the milk. Then they went back east to Arkansas and bought a couple of creameries back there. And they condensed back there, and it wasn't very long till they had more condensed goat milk than what they could sell. They wanted to cut us off for a while, because they could produce it cheaper there. I told them right away I didn't like it. But they said, "You don't need to worry. They're Grade B and you're Grade A." I said, "Well, you're using that milk to take care of the East. Doesn't make any difference whether it's Grade B or Grade A." And they admitted it. Wasn't any time at all till it was like I said, and they told us to get out of the business.

It really pretty near ruined George's life, I think. He'd get up at three or four in the morning and come over and milk. I wouldn't get there until about five or six, but I'd help. It's harder to get ten gallons of goat milk than of cow's milk, and we generally milked 200 head. Goat milk isn't like cows' milk, it's more perishable. They have to turn the cans every 30 days or it will settle and spoil.

I went into chickens after George did — I wound up with about 15,000 after five years. My youngest boy, all he could think about was hogs, and when he let me know that he didn't want to fool with chickens, why, I just got rid of them too. By that time I had some almonds on a ten-acre piece I'd added. I really did better after I got almonds than any other thing I did. But I made some money on the chickens, to tide over on the bad years anyway. Sometimes we got down below cost, maybe because of overproduction, or maybe the experts would come out and say that eggs are high-cholesterol and scare people so they'd stop eating them for a while. Generally eggs are up certain times a year and down certain times a year, but there's some years that it gets down too cheap.

Up until I quit, I belonged to Colby Poultry Producers here in Beale, a cooperative. They went under. Then I sold to Jenkins, a private guy. I did all right with him, but when Peter didn't want anything to do with eggs, I quit and rented my buildings to George for a year or two. Then my son-in-law thought he'd go in with me, and we bought a bunch of chickens, but boy, we bought them in a bad year. The egg price was down; that soured him. We got a contract with Alberts, but it was nothing but disappointment all the way. We stayed with it till time to molt the chickens. After we molted them we made a little money.

I used to tell George to molt his chickens, it'd pay him, but no, he'd only keep them 18 or 20 months and sell them and get a new batch. I got started on molting — that's when they lose all their feathers. You just take them off of feed for a certain length of time. You have to give them a little water after so many days. It takes, roughly speaking, two months to molt them, six weeks at least. And when the feathers come back in, they lay good eggs again. They don't lay while molting, of course, you don't want them to. If they're laying a few eggs, you aren't hitting them hard enough. You want to hit them hard enough till they get poor and have to fatten up again. Then they come back in and lay a good egg and you don't have any little eggs to deal with. They're older hens. I made more money by molting than I did buying pullets. Chickens can be molted even a third time. They would be two years old to have a year laying, and then you'd molt them, and molting and laying would be another year. If you did that a third time, you figure you might get five years from a hen, maybe.

I tore my buildings down after my son-in-law quit. He'd had enough trouble and was always worried. He thought we was going to lose everything. Had to give Alberts a second mortgage on our place to get him to finance us. But we finally got out from under.

I sold the 20 acres about five years ago. It has a hundred houses on it now. The farms that are close to town, they began to sell for subdivision. People complain that all this good farmland is going into houses. Well, it's a shame, but on the other hand, water is going to new locations today, and money is developing farms by the square mile. Little farmers are in competition with great big operators. I think it's a good thing when they can get out with a little money.

But land is going so high now, I don't know when it'll end. When I sold my place the market price then was $7,000 an acre. I could get $21,000 now. There has to be some limit to it, I would say. Of course, we're right on the edge of Modesto. The house next to me is in town, while I'm out in the country — the line goes right between these two houses. I've got about two acres here with 80 walnut trees, and it's a nice place to be retired on.

My sons are acquainted with finance, they have to be. I didn't know anything about finance. If you happened to be fortunate and raise good crops and hit a high price, well, you did all right. Or you could start an undertaking and happen to hit it at the wrong time and maybe it wouldn't be your fault at all. The saying used to be, "Don't put all your eggs in one basket." You had a few hogs, a few chickens, a garden, a few cows. Now if you're going to run a dairy, you better specialize in it, because that's what you're in competition with. But when I was a boy, that was one of the things they told you, "Have a little of everything."

When I was starting, if I had a team, a scraper, and a plow, I could rent a place and get started farming. Today you have to have enough to retire on if you want to get started. My boy back east had trouble with his harvester and thought he'd buy a new one, a little bigger than what he had — and it was $90,000. "Why," I said, "when I was a boy, if I'd had $90,000, I wouldn't stick it in something like that. I'd quit — I'd be rich!" That's just the difference. Boys who would like to farm today haven't got the money. When I was young, it didn't take money, it took work. Had to learn how to run water uphill and all that. Now it's bigger machinery and money.

## ELIZABETH LOHMAN LOWE, AGE 71

I was born in southern California in 1907. My father was a carpenter. During World War I we lived in Long Beach and he worked in the shipyards. My folks owned a house and lot on Signal Hill. At that time the Japanese people raised beautiful gardens there, and we'd go up and get cucumbers. We sold the house and moved here in 1919 because my father had heart trouble. Down in Long Beach it was foggy and damp and he thought he'd be better off up here. They discovered oil not long after that on Signal Hill, and the people who then had our property really cleaned up on oil. I'm glad somebody got to.

I'm the oldest of 19 children. We would have a big skillet of fried potatoes and that would be supper, or a big pot of soup. I remember when we thought a loaf of bakery bread was better than cake. My father always thought the pasture across the fence was greener. We lived in Whittier where he worked in the citrus groves, and we lived in Long Beach, and in different places for just a year or two, and then we'd move someplace else.

Up here in the valley we had a farm, but my father was really no farmer. When we first came my father raised Gyp corn

— like milo, only it's white. He had a few cows, a few chickens. Even in town in those days everybody would have a little coop out in back and have a dozen or so chickens. That was just about all the meat that we had. Daddy would always have a nice garden. We never went hungry, but lots of times we sure would have liked to have had something different than we did. But it didn't hurt us, really. We all grew up healthy. My brothers, most of them, live in the Bay Area. They all have good jobs and did real well in life.

We had to walk two and a half miles to school. The children would all start out and the oldest brother would carry the youngest one part of the way. I'd stay home and get part of the work done. Then I'd have to hurry to get to school. I only went through the eighth grade. I just about bawled my eyes out when I couldn't go to high school. I wanted to go so bad, but it was impossible. We lived about 13 miles from the high school, and I just physically couldn't get there. There were no buses then. They say that if you want to do something bad enough, you can. But even now when I look back, I don't see how I could have, because Mom was either always pregnant or else we had a baby. Even if I could have gone to high school, I wouldn't have had clothes and things so that I could have enjoyed it. So I didn't go. The younger ones did, but I didn't.

When I was 14 my folks decided to move to San Luis Obispo, so I went to work in town here as a house girl. My brother went to Hawaii in 1926 with some company and dredged out Pearl Harbor. During the summer I worked in the cannery, and one year I worked in the labeling department all winter. I was 19 when we were married. I had met Otis through the Church. My best friend was a member, and I'd go with her to the young folks' parties, and that is how I met him. After we were married we joined the Church ourselves.

I never helped with the dairy work. I did help gather eggs some, but not regularly. The house and children kept me busy, so I didn't do much farm work. But I did work off and on for outside money. During World War II I worked at the dehydrator, processing dehydrated carrots for overseas. Just a farmer started it up; then he got bigger and went to assembly line, and I became forelady of the swing shift. This worked out real well, because Otis was home then with the children. That was year-round work. They hauled carrots from Bakersfield and from all over. I forget how many tons we did a day, a huge amount, all for the Army. We got an "E" for excellence in production. I was one of the supervisors.

After the dehydrator closed down, I had charge of the grape packing shed. That would usually start maybe the first of September and go into November, packing grapes and shipping them east. During the Depression a place across the road from us grew sweet potatoes, and I used to sort them. Then every year I'd work on the almond hullers or in the packing sheds. I always was doing something for outside money, trying to make a contribution to the family income.

One reason I started at the dehydrator was because our house, and everything that we had, burnt. We'd just bought the place. We had no clothes, no food, no money, no nothing. People were really good to us. They brought things in — our people do things like that. But the children needed coats and warm clothes, so there was nothing for me to do but go to work.

I cooked for several years over at Santa Cruz for youth groups and church camps. But I never left the children without Otis being there. That's one of the problems with children today, they are not supervised and disciplined. We thought the children should be disciplined as well as loved. We always gave them plenty of responsibility. Later I went to junior college and took some classes in nutrition. I was not a registered dietician, because I didn't go through a college program, but I did become a dietician at the convalescent hospital for about five years.

In our sons' families the daughters-in-law don't work out. They're all at home with their families, which is the type of life that we encourage. I had to work out because of the Depression and so little money. But it's better that a woman doesn't.

George Lowe is a small, neat man with a salt and pepper beard. He is positive, righteous, sure of his worth and his values. Articulate as a preacher should be, he can describe the conflicts he feels as a businessman and as a moralist. Farming and church work are the two poles in his life. His wife, white cap tied under her chin, hair wound in a bun on her neck, is clearly a member of a religious sect. Her reddish dress is of an old-fashioned cut, capelet attached to a peter pan collar. Her face is plain but soft and pleasant, her manner dutiful.

GEORGE LOWE, AGE 50

I was born in Modesto in 1928. My grandfather came to this area from Virginia. Part of his family went to Southern California before the turn of the century and part of them came up here. They had the pioneer spirit and the wanderlust like a good many people did at that time. They were seeking financial opportunities. My father was the first boy born in the Wood Colony. That land was subdivided when irrigation came in. I remember my grandfather very clearly. He bought land out on

Backus Road near here. When the irrigation came in, the alkali came up (they were watering before the pumps came in). So he had some financial setbacks, because black alkali would kill the alfalfa. He went to other towns for a while but came back to this area.

As for myself, I went through the local schools. My father suffered setbacks during the Depression. After I was old enough to be major help on the farm, he bought a place in 1941, real close to here. We milked cows through 1946, and goats for a while. I have a background of very hard work. I couldn't stay awake in class at school because I'd get up at four o'clock and milk in the morning.

I have three younger brothers and a sister. Two brothers are in Ohio now. One is a full-time farmer, the other owns some land and works in construction. Another brother is an engineer for the city, and my brother-in-law works for the postal service. I think I could have done a number of other things, but I fell into farming and never very seriously considered anything else. A time or two when things went adverse for me I'd wonder if I shouldn't be doing something else. I was never in the military service nor did I go on to college or any other training. I really started farming right after high school, so I've been living in the dirt all my life!

My wife and I moved into the house we now live in, on the night our twins were born. We've lived here all that time. I was farming with my father until 1956, when I was 28. Then I went on my own. All the time I was growing up my father told us boys, "If you'll stay with me and work on the ranch, I'll give you 20 acres, and that will be sufficient for your living." This was more than my father could do, really, even though his intentions were good. He is a fine man whom I respect, but financial success never really came while we were growing up.

The interesting thing is that then we thought 20 acres would make a living — all through the 1940s we thought that. Even in the beginning of the 1950s. When I was a boy, every 20 acres had a dairy barn on it with 20 to 40 cows. There's still old dairy barns around, there's one across the road yet, but most of them are gone now. In those days if you wanted to borrow money at the bank you had to smell a little bit like the cow corral. When you walked in there, your chances of getting money were better if you had a little manure on your boots, so to speak. The dairy farmers were hard-working and honest, with a paycheck coming in regular. They were a good loan risk for any business around, including a bank. Well, the dairy business has evolved so that there are just huge dairies now — no small ones left. We used to sell milk in ten-gallon cans, and would skim the cream off the top to make ice cream — my memories of real rich ice cream are delightful. But the pressures on these small 20-acre ranches were just too great.

One thing that squeezed out the small dairies was the upgraded standards. It cost so much money to upgrade to meet Grade A standards that the dairyman had to have a lot of cows. Then, as small dairymen dropped out, there ceased to be even a market for the small supplier of milk. With the extra handling, the margins became so slim that it was not just a matter of poorer living — the small dairy couldn't make it at all. Economic pressures have driven them all out. A few years ago some around here still shipped Grade B milk to Hershey for chocolate, but I don't know of any now. As the Grade B dairies were phasing out it was often the older farmers who retired, but it also cut across age groups. I know young men who wanted to farm, but it was economically impossible. I almost ruined my health trying to make our goat dairy go, and I have a friend joining me on the back of my place, hard-working and honest, who just wouldn't give up. He was young and he almost ruined his health and his financial status, just trying to make a dairy go, because that's what he'd grown up with. But it simply was not profitable. The big suppliers and the Grade A contracts were where the money went.

I milked 200 goats myself for a while right here. Then in 1956 that business went very sour. They told us in September that they wouldn't take any more milk that year, and they gave us a quota for the next year of 80,000 pounds, or about a fourth of what we would have shipped. We'd spent many years building up a fine herd, much better-quality animals than we'd ever had in all the years we'd been milking. Little kid goats, doe kids, that cost us $20 to raise, we had to sell for $10. Beautiful animals. Everybody else was in the same boat. It was a real catastrophe.

Goat milk is too small a commodity, so the marketing just was not steady. They can't hold evaporated goats' milk as long as they can cows'. It gets strong. When they had too much goat milk all they could say was, "We won't take it." I don't know why they got too much goat milk just then. Perhaps the market shrank some. A lot of the goat milk had been for babies, and some of the new soya milk and other types of baby foods came in around then. Goat milk was sold at the drug counter rather than in the regular food stores, so it was expensive. The marketing channels weren't ideal. There are very few goat dairies left in California now. It would be nearly impossible for anybody to start with a few goats now and try to make something out of it.

At that time, 1956, I decided to split off from my father and go my own way farming. So I put poultry in. Those were our years of financial struggle. But ever since then I've stayed with the egg business.

I began to realize, as I got into poultry, I had to get bigger. Once I thought that if I could have 10,000 birds, that's all I'd want. Today I have six times that many. In order to stay in the full-time business of farming, it's been necessary to increase volume much more than we ever would have anticipated even 20 years ago. There was double pressure: First, new facilities became available, and the rewards were more attractive with a larger operation. One man can take care of even 50,000 or 60,000 birds today, if he's got a fully automated house. Earlier, that would have been unheard of. You could only handle 3,000 or 5,000, maybe, before things were automated. Then the pressure for larger business grew as other businesses around got bigger. For instance, bulk: We haven't handled sacks for feed for many, many years. We decided we had to buy feed in bulk, not sacks. You had to have 3,000 or 4,000 birds even to go bulk. Our feed costs would have been tremendously high for anything less than several thousand chickens. Today you have increased feed costs and delivery costs for anything less than 20,000 or 30,000. These pressures are great. Even with 60,000 birds I'm one of the smaller egg suppliers. I am just an egg factory, because I don't do any packing or cleaning. Suppliers don't want to send their delivery trucks out for a little bitty operation. Most of the ranches are 100,000 birds. If you have just 5,000 you don't have any economic pull. They could add 5,000 more chickens somewhere else and do without you easy. You have to be a certain size even to keep your name on the rolls.

But there is more than that. Used to be, maybe you could make a couple of dollars a bird, and that would have been enough to live on. But our return per bird has actually dropped over a period of time. When 50 cents or 25 cents a bird is your profit, you have to sell more to make the same total — and then we require more money to live on today, too. All this has made real honest economic pressure for larger numbers.

I don't believe that any conspiracy was going on. It was just an economic fact of life. It was change in our society; you can't fight change. The loss of these small farms, though, has been one of the things that has contributed to moral decay in this country. I'm not a pessimist when I say that, I recognize good in the world too. I just read the editorial in the last <u>US News and World Report,</u> and that writer is more pessimistic than I'd be. I'm not reflecting some oddball religious idea that everybody's horrible — I don't think that way. But crime and moral decay have definitely been accelerated by the demise of the small farm and the small business. This is widely understood in our society, I think, by those who study it.

A boy who grows up from the time he is five or six years old, taking care of milking the cows, getting the eggs and taking care of the animals, and knowing that if he doesn't do it, the family will suffer, learns two major things. First he learns responsibility. He's got to take care of that animal or it starves. Second, he is an integral part of the family, and he is needed. We grew up knowing that if we didn't work, we didn't eat. My folks didn't manufacture jobs for me to do. There was more work than could possibly get done. I've had the same experience with my boys. I didn't bawl my older boy out for breaking a few eggs, he cried because he broke them. My younger boy, now, has come to realize that Dad could hire somebody else if he didn't do the work, and that has not had as desirable an effect on him. Of course, we like better times, and I don't wish hard times on anybody. But they obviously did have a sobering effect on young people.

I did some welding for outside pay at one time when times were rough. I don't have to do that now. One of my boys works steady with me, and I also have a steady man. The other boy works part time and owns one piece of ground with me. 1956 through 1958 were the low points for me. All through the 1960s I was on the way up; I was putting in more birds. I have two separate chicken houses now. I lease a 40,000-capacity house and have the management of that leased out. Then I have 20,000 right here on my own place, in a house which I built, and my son and I handle that. Some school children gather eggs here every day. I use youngsters to give them an opportunity to work. Then I farm about 200 acres of almonds. About half of those I own, the other half I lease.

I feed my own 20,000 birds on this ranch night and morning. Takes me half an hour actually feeding, but till I count the eggs and do some little fixing, it is a couple of hours. I help the boys at harvest time. I'm a good fill-in man, a repair man, whenever there's work that needs to be done. But in winter my days are really kind of free. I get up at six every morning and read the paper, then go out at eight and feed my birds. An office does my bookwork and accounting. I'm not as hard pushed as I used to be. I have stepped down in the last few years of work. I've backed off from opportunities to get bigger for personal reasons.

In the past I farmed ground almost all the way into Modesto. The town is right at my back door now. The farms are going to housing and urban growth. Economic conditions have changed and standards of living are higher for everyone now. I wouldn't wish my boys to work today like I did at their age. Nobody works that way now. They won't stay on the farm milking cows if they're losing money. They can get a job somewhere else. But we milked cows whether we made money or lost it. I used to argue with my Dad about that very thing. But my Dad was a little different from everyone else around. He was willing to take a loss just to stay in the business. He didn't know anything else, and he figured sooner or later it would come out right. He was an eternal optimist. Actually it did come out all right for him in the end.

In the last two years, ten eating places have opened on the intersection less than a mile south of me. The place across the road was sold two years ago to a subdivider. He's just waiting till the time's right to build more houses. The reason I've been able to stay here is because with 20,000 birds on 15 acres, this place is highly productive. My 15 acres here have tripled in value in three years. The man who bought that piece across the road paid about $6,000 for 50 acres, and now it's worth $20,000 in two years' time.

This is not right. I'm concerned about what inflation is doing to the poor man. What happens is that anybody who owns property or a thriving business has just greatly increased his holdings. He's a much richer man, even with inflation considered. But the poor man with nothing is that much worse off. It's going to force more people on welfare. And when you start to give a man a dollar, you take away his personal dignity. You make him a beggar instead of a contributor.

The man who works for me now has been with me for over two years, and I've become very well acquainted with him and his family. He is a Mexican-American. He had an alcohol problem to the point that he left me for a month at one time — just got to where he couldn't even work, and he finally ran off. He lost his wife and ran up bills. But he went to A.A. and has now been dry for a year and a half. Now he's 40 years old but he has nothing. He's competent, intelligent, very likable. He's got a sense of humor. When I give him a shovel, he says, "Ah, Mexican bulldozer, tamale power," and away he goes. He's a real fine fellow, but he has just wasted his money. He doesn't know how to budget. He buys a car and pays too much for it, and it just goes on and on.

Employers are always looking for good employees. The real problem is the lack of moral values in some workers — and a concern for the value of the dollar. Their opportunities are immense. If a man were willing to save a dollar or two and live right, there's still a possibility for even a man at the bottom of the scale to gain, providing he has reasonably good health.

My man speaks good English, he's bilingual and is very strong in both languages. He isn't highly educated. But at least he's doing very well now, and he's happy too. He's a missionary for A.A. He's going to get along real well if he stays with it. His children were raised while he was an alcoholic, and he and his wife weren't getting along too well either, so the children are having problems.

Sun Valley Farms are where my eggs go. They're the fifteenth largest egg producer in the United States today, with a million and a half birds contracted. They have a very fine company. I benefit by their expertise. They grow my pullets, and package my eggs, which are sold as Nulaid-brand eggs. Good operations, all of them. Very, very competent men.

As farming gets more technical, someone who makes the right decisions can make a profit. There's no need for the farmer with management ability to handle a hoe or shovel, or irrigate. But years ago, a man who had 20 acres did his own tax work, his own calculating. He was everything — the laborer, the marketer, the irrigator, the milker, the tax expert. He did it all. Today we're in an age of experts and concentrated ability.

Take eggs: One hen on an average in a year lays 22 dozen eggs. If through careful management a man could make that hen lay eggs for 2 cents a dozen less than another man, that would be 44 cents per hen more money that he made. Right now, today, large eggs are 60 cents a dozen to me wholesale. Now 2 cents a dozen is less than 5 percent, yet 44 cents per hen on a 60,000-bird operation like mine is $30,000 a year. When we're dealing with large numbers, small advantages can amount to an awful lot of dollars.

Back during the 1960s we were increasing on the average of two eggs per hen per year, as a rule of thumb. It was amazingly high. In 10 years we were producing 20 eggs more per bird, through a combination of better housing and management practices, of better breeding and of better feeding. This progress is still continuing — in the last three weeks I've had the best egg production here that I have ever had. I'm breaking records. I recall one time years ago a feed manufacturer said that they had reached the ultimate in feed, and it was up to the geneticists to improve production. He was laughed to scorn, really. We're still making progress in feed as well as in genetics.

Four major factors are important in egg production: number of eggs (you've got to have numbers or you don't have anything to sell); size of eggs; the amount of feed consumed; and eggshell quality. The interesting part is that if an egg geneticist or breeding company got a bird that would lay larger eggs, you could almost depend on it that the shells wouldn't be as good, or else the number would go down. To get an increased number of eggs, you'd get a decreased size. Almost like adding one and one is two. We compare breeds of birds to replace our flocks with (they're all Leghorns, but they're from different breeders). We buy from five major breeders today. We play one factor against the other — the companies are strong on different factors. It's a real challenge to them to get the combination that will be the most profitable.

In egg size, we have extra large and jumbo, then large, medium, and small. Some companies pay more for extra large and jumbos, but generally there is more breakage in those big eggs. If a hen could lay all large eggs, and no mediums or smalls, and no jumbos or extra large, we would make more money. So you shoot for birds that get into the large size as fast as possible and then don't get too big too fast.

We have the most rigid standards that you could imagine. We throw away tons of food in America because our markets won't take anything that's not prime. Farmers just literally cry at the waste. When I started out with 1,300 hens, we'd bring the soft-shell eggs in. (A hen will sometimes lay eggs that don't have a hard shell, just the membrane around them.) They're perfectly good if you bring them right in, but if they lie in a cage one day the moisture evaporates out of them and they become all leathery. But they're just as good eggs as the rest of them. We used to eat them. For years we'd ship the cracked eggs in to the company, and they would break them and use them for noodles or whatever. But today generally if an egg is noticeably cracked here on the ranch, we just throw it under the cage as manure.

America would be better off if we didn't have so many large farms. But there's constant pressure on farmers to get bigger. One of my friends now is moving to Pasco, Washington, where there's room for a bigger farm. I'm not rubbing shoulders with corporations here, so I don't feel them as an immediate threat, only in that if they can produce almonds more economically than I can, they have a bad effect on me. But we have an advantage here locally. We have excellent water and soils, and a fine climate for all kinds of crops, so small farmers here have about all the advantages, as well as good cooperatives marketing for them.

Actually, there's still an opportunity for small farmers that's overlooked. This would be a local retail business for eggs. You need at least 20,000 chickens to be counted as a commercial layer operation today. But a person who wanted to put 3,000 birds in and retail his eggs could spend half his time in marketing and half his time in production. There is a market for roadside sales. The demand is terrific for fresh eggs. People in town would prefer to purchase fresh eggs from the farmer. They ring our doorbell all the time, but our contract says that we can't retail eggs, and selling also takes time. A small farmer, though, could have a special stand along the road, or build up his own market with deliveries to institutions and restaurants. If someone with enough ability chose to do that, it could work. But as far as contract sales to bigger middlemen, a small operator's time would be worth so much more somewhere else that it wouldn't pay him at all.

I sold a piece of property two years ago, and when I rebought I gave 40 acres to each of my boys. They each have a third, and I have a third. I did this specifically so they could have some hedge against inflation and some incentive. My boys are conservative. My older one in particular is ultraconservative. He's very competent but he would rather not be so obligated to farm work that he can't enjoy himself. My second boy is just getting married. How he'll develop I don't know. The two of them may work for me as employees for a while and phase more into partnership. They have stayed with me, they like farming. I won't be here forever. Then they'll have whatever I have, sometime. My older boy has taken over the management to a large extent, because my church obligations are greater all the time.

We belong to an agrarian church, and we are facing tremendous urban pressures. When I was growing up, nearly all my classmates worked on farms, as well as my church friends. Today very few of my peers do. Out of a congregation of 300, I could count on two hands, maybe on one, all the real full-time farmers that are left — and that's in a church committed to an agrarian way of life. The major portion of our people now are in the building trades. They build houses, do masonry, plumbing, electrical work. They've been very successful in our area because they're very competent, honest people.

The church is central to our lives. The name of our church is Old German Baptist Brethren. We're part of the Brethren group that came over to America in 1719, part of the German movement into the Penn Colony. In Revolutionary War days our members printed the German Bible in America. We are a "Plain People." While we grew large at one time when America was growing during the nineteenth century, Plain People are not numerous in our country today. One historian says we're one of the few Plain groups that still shows modest growth. We came out of Germany from the area where the Mennonites also were, but that's just about the sum of our formal relationship.

All Plain Peoples have three major tenets that are different from other Christian sects. We believe in separatism. This is the reason for our dress. "The world passes away," and less thereof is better. We think that if a person is really an honest Christian in his heart, he should look like one. My uniform is different from the clothing of other men, my trousers a little different cut. I wear a straight collar, not a tie. The women all wear a cap called the "prayer covering" most of the time. We feel a woman who prays or prophesizes with her head uncovered "dishonoreth her head," as the Book says. This is a very obvious deduction if you study the Eleventh Chapter, First Corinthians. All Christian ladies used to wear a hat or a head covering of some kind. Beards are mandatory for those whose responsibility is leadership and widely worn by other members, but not mandatory.

We also believe in the Bible's inspired word, and we believe in keeping the Commandments as they're given. We also believe in Grace. Sometimes people think we get salvation by works, which is absurd.

Most all Plain People have a central conference with a certain amount of jurisdiction over their members. This is extremely unpopular in America today, because nobody wants anybody to tell them what they can do. We do not allow members to own television or radio; they are not of help to the proper raising of our families. Most people understand this even if they're not sympathetic otherwise. There is a very strong emphasis on the family unit in our church, and on youth. A big group of our youngsters went ice skating night before last. Our young people have something to do every night if they want to.

Our local congregation is the largest one of our fraternity. We have well over 300 members. With four churches in the county, there are between 700 and 800 of us locally. This is one of the major gatherings on the West Coast.

RUTH LOWE, AGE 46

I was born in Michigan in 1932. My father farmed, then he quit farming and worked in a factory. He didn't like the snow so we moved to Ohio, and we lived there for about six years. He didn't like the snow still, so we moved to California. My uncle had moved here previously, and the Church was here. We lived just down the road and I went through the local schools. I didn't graduate. I quit after my junior year to get married.

Our first year of marriage started with a bang. We went east and bought a new car and drove it out of the factory, and that paid for our trip. Then when we got home we started to build this house where we live now. Then, ten months after we were married, our twins were born, a boy and a girl.

George was milking goats here at that time. About 1956 the business went broke. The wholesalers wouldn't take any more milk, so we had to sell out. He thought about trucking, but I didn't really want him to do that, and he wanted to go into chickens anyway. He had these barns available, and chickens were something that he thought would work into the situation. It was a gamble, but he did it. We liked the chickens, but our living was modest, and we were just getting by. Not quite three years after the twins were born we had a little boy. Five years later Joanne was born.

We had a little egg cleaner and I cleaned the eggs for about an hour every evening. As we enlarged the chicken flock and had more eggs, then that time grew longer. Sometimes the children would help take the eggs off from the other end. We put them in cases for the egg man to pick up. The children put them on the flats, 12 flats in a case. When we started with eggs, Andrew and Alice were in kindergarten. Alice always helped me in the house, so usually just Andrew would help me in the egg room. When Paul got old enough, he gathered eggs, too, and so did Joanne. The children helped every day in that way. It was good experience for them. In the house you can put off folding clothes or other things, but you can't ever put off gathering eggs. It had to be done every day.

I hardly ever had a garden. We maybe had a couple of tomato vines or squash, but usually the Bermuda grass would take over. I just wasn't interested in it particularly, I'm not a gardener at all. I tried to do the bills at first, but as things became more complicated, George began using the secretary over at Biggle's office. It was easier. He doesn't have enough to keep a full-time secretary busy, and this way she writes all the checks and pays all the bills. I am not very smart at arithmetic and I don't know anything complicated. I can write a check, but as far as figuring anything out, I can't. Percentages and all that kind of thing I don't know. Arithmetic was not my subject.

I leave things completely up to George, because when we first got started, I worried about the bills, and it wasn't good for my low blood sugar. I get along lots better if I leave the decisions all to my husband. George never had an ulcer, but he had to watch his stomach. It was worry and bills and things when he worked awfully long days. Farmers always worry. Back when he was milking goats he got up and started at 3:30 in the morning, and he would work till dark if there was field work to be done. You know that saying, a farmer's work is never done.

George is a minister in our Church. In our Church the deacons and the ministers are elected. Ministers are in the first degree for about six years, and then they go into the second degree for six years. Then they're made an elder. The first-degree minister more or less studies; the second-degree minister baptizes and marries. They all take their turn preaching. Our church is a multiple ministry. In 1962 George was elected deacon, and about 1964 he was elected minister. He is an elder now. That is life-long. We feel that the spirit has elected you and you're a minister for life.

George preaches regularly. We have church every Sunday morning and every Sunday night. The ministers sit at a table in the front of the church and the deacons sit at the opposite side. They start all hymns and they read the scripture, the morning lesson. So even if George doesn't take the text, he still has to be prepared for opening and closing.

Of course, no woman speaks in our church. The ministers' wives sit in the first row and the deacons' wives in the second, with the congregation in back. The ministers' wives take care of the tablecloths and the wine bottles for the Communion. If somebody joins the Church, the ministers' wives get a pattern for them. I am a seamstress, and one of my gifts is to do patterns, which is hard. Since we make our dresses all the same, a pattern is quite important, because if you don't have a pattern that fits, then none of your dresses will fit.

Our people were originally mostly from farm families. Now a lot of people can't afford to farm, to buy the land or the equipment. In Ohio, Indiana, and Pennsylvania the Church still has plenty of farmers, but not out here. But it doesn't mean the end of our Church just because everybody can't farm anymore. There are other occupations. We have doctors and engineers. I don't think we have any lawyers, but there are carpenters, plumbers, and electricians.

Our children have all stayed in the Church. They see a trend toward liberalism and it turns them more conservative. Our son Andrew is even stricter than what we brought him up to be. He's got a mind of his own. And our daughter is more strict; her little girl wears dresses longer than I made her wear hers.

Our social life completely revolves around the Church. We do not believe in belonging to political groups or to secret clubs or the Lions or anything like that. We do belong to associations for business. But we don't take part in politics. We don't even vote. We feel that we can do more for our country by prayer. When you're in politics, it involves too many things. And we feel as in the Bible, that our kingdom is not of this world. In the Bible, the Lord taught us not to fight, so if you don't fight, you shouldn't vote. We think politics are corrupting. Our young men are all conscientious objectors.

Our boys grew up with George in the fields. They always went with him. When the father works away from home, boys don't have that close association with their father. And when children are in town, in the street with other children, parents don't have full control. But on the farm children learn responsibility, and they don't have that other association to fight all the time.

As the third generation in this farm family, Andrew represents an evolution in attitudes. The uncertainties of 75-year-old Otis are smoothed out in George, who has succeeded where his father had not. Confident as George is now, he still can weigh the differences between the past and the present, and express a sense of loss for some of the intangibles of a simpler way of life. But young

Andrew, with more future ahead of him than past, is supremely sure of himself. His religious commitment is reflected in his beard and his plain-cut clothes, and his bright blue eyes show no trace of doubt. He expects the future to be good to him — he is in the right place at the right time.

ANDREW LOWE, AGE 28

Religion has had more influence on my life than farming or anything else. My memories are that farming was a struggle for Dad. We didn't really have any money at all. When they turned down the goat milk, Dad had to sell part of his farm in order to keep what he did have. So many memories of childhood aren't of being well off at all. But we were always a close family. I grew up thinking my Dad was perfect. As far as I knew, we had everything we needed.

I'm the oldest, a twin. From the time I was in kindergarten, I always gathered eggs or had other chores. When my sister was born, I was seven. Dad had two barns of chickens then and I would go out to help feed them in the morning. I would push the cart. When I had to push it through mud, I couldn't hardly do it, but I did feel that I was making a contribution.

I won quite a few scholarships in high school, mostly awards for FFA, and in order to use them I went on to junior college, where I took some poultry classes. For Future Farmers I'd had one of my Dad's barns of chickens as my project. It was right after a big depression in egg prices, but prices went real high again, and that made the project a success. I always had a high academic record, too. My teachers put pressure on me to get out of farming and do something more academically inclined. I was always good in school. If it wasn't for my religion, I don't think I'd be in agriculture today.

After the one year of college I didn't go on, mostly because the junior college courses were what I knew already. I never even considered going on to college elsewhere. I thought I'd like a family farm where I could have a welding shop. My father had the expectation that I would stay on farming with him, and I had a girl friend. Now I'm an employee of my father on the egg operation. I'm on salary. But I'm also in partnership with him on the young almond trees we just planted. Several years back, maybe three-fourths of our farm went into houses. Dad sold his ground before prices really went up, and he couldn't buy anything soon enough to gain what he could have, but he did well enough. So we started buying other land by the river and have planted it to almonds.

We've started a new side business just recently. We farmed our neighbor's piece back of us, and the only way we could spray it was for me to get a license. I got one, and we decided maybe we could do more spraying on a commercial basis. So I built a special boom for our tractor, and now I'm a pesticide applicator. It's called "Lowe Farm Strip Spray." We're not exactly sure how we'll work it out as far as cash goes. All the equipment is my Dad's. If it works out to be my business, there will have to be a rent agreement. The way it is now, I'm still working for him.

I was married when I was 20. My wife doesn't have a farm background, her father's a carpenter. But she had a rural base. We have three small daughters. Right now we're renting our house from a friend, to be close to Dad's place down the road.

We have more than our parents had when they were young, which affects the way we live and think. Dad never had the base to work from which I have. He had to be aggressive to recover from the dairy goat disaster. He stuck his neck out a few times and worked much harder than I do. If I have a little trouble with my sprayer rig, I have some ten- and twelve-hour days. But usually I try not to work longer than eight or nine hours. And I never ever work on Sundays. At my job, according to my salary, I'm supposed to put in a 45-hour week. It averages out to that. Things are so different now, so much more mechanical, that the actual real hard physical work my Dad did, or even that I did as a boy working with him — hoeing around trees, chopping weeds and things like that — have been minimized tremendously. Nobody needs to do that anymore.

Independence and farming go together. I like to be free, to be my own boss. Some people think being close to the soil is being more in contact with God, and I feel that way too. I grew up on the farm and Dad was always around. If he went somewhere, I went too. We all did things together. That's my most special memory. I drove a tractor from the time I was maybe five or six, young enough so that my Mom would worry. I'd like to keep hold of that for my family.

I'm very optimistic about farming. I think we can go about as far as we want. Of course, we don't know what future land prices will be. We want to stay in this religious district. We can't buy $5,000 ground that's 80 miles away, because we're in a religious community. We have to buy $10,000 land around here in order to stay. With prices going up like they have, our whole attitude toward land has changed. The chickens have been a very intensive use of land. We can operate a big chicken house on a very small piece.

The town will stop growing sooner or later. I've seen more no-growth attitudes from my days in school than some of the older people realize. Whether it's possible to keep that when the dollars start flowing around, I don't know. But there are more ecological types around, with strong feelings about good farmland being covered up with houses.

A 1918 photo shows the "old way" of milking cows, complete with dairymaid. Reprinted with permission of the San Joaquin County Historical Museum.

# PREVAILING WINDS
○
## The Schoppe Family

Walter, the patriarch, is a sturdy German-American, solid as an oak at 75. He paid his dues in the Depression and has weathered many a storm, and in his age he is accorded the respect befitting a successful survivor. He sits in the spacious kitchen of his comfortable farm home while activity swirls around him -- ringing telephones, grandsons working in the yard, Cornelia taking cookies out of the oven. His world is secure. He has struggled to make it that way, a hearty man who has met life head-on. His attitude toward Cornelia is protective. She too is diligent and careful, but frailer than he.

Their only son Bob is their pride and joy: handsome, intelligent, an industrious farmer, and devoted family man. In faded plaid shirt and jeans, boots muddy from the rain-soaked fields, his genial air belies the fact that he is constantly juggling the myriad of details necessary to the smooth operation of a highly diversified farming operation. His wife Gwen is a vivacious, pretty blond, full of enthusiasm. One can imagine her surrounded by a crowd of admiring kindergartners.

WALTER SCHOPPE, AGE 75

I was born in 1903 near Stockton. My dad had settled on Roberts Island in the 1880s. He came over from Germany when he was 16 years of age and worked on a ranch where his sister had married. Ten years later my mother at age 15 came from Germany to San Francisco and lived with cousins there. The folks were married in the early 1890s. Dad rented a piece of property 12 miles from Stockton in the Delta and started farming. After living on that property for 18 years they bought a larger place nearer to Stockton and then farmed both places.

We were a family of eight children and all of us were born on the farm. When we moved closer to Stockton, my youngest sister had the luxury of being attended at birth by a doctor. The older boys went to work right out of grammar school. The number three son had the opportunity to go to Heald Business College. Number four got in three years of high school. I got in three years of high school and then it was necessary to work on the farm. I had three younger sisters. The oldest graduated from high school, did some secretarial work, and later married. The next one went into nurses' training. The youngest went on to college and got a teachers' credential.

When I was about 16 I started working for my father. Not all his sons stayed with him. This was a grain farm and there just wasn't that much activity. The oldest brother went to work for Holt Manufacturing Company, and the others started farming for themselves. They bought tractors and did custom work. After I worked on the home ranch for a few years, I went to work on a ranch in Santa Clara Valley.

It was through that ranch that I was introduced to the Davis campus on a short course. They had a deciduous fruit short course there by Dr. Robbins, Warren Tufts, and Dr. Howard. I spent a delightful week there in 1922 and that opened a whole new world to me. I was very grateful for the opportunity, and starting in 1923 I put in two years on a nondegree agricultural curriculum. I went to work as a blight control operator on a pear orchard near Farmington, and a year later became foreman. That ranch was sold two years later, and I rented this ranch for myself. Two years later Cornelia and I purchased it on a rather generous contract. So the Depression finally created a great opportunity for me, even though I was seven years out of college.

When I rented this property it had about 110 acres of peaches, about 30 of apricots, and some open field ground. The owner had lost money on it for a number of years. I made a proposal that I would rent the thing on the basis of 50 percent of net profit, or 50 percent of the losses. (Obviously I couldn't have absorbed much loss.) The thing that appealed to him was that if he had no profit, he had a foreman who worked for nothing, and if there were a profit, he was willing to give up half of it. The proposal was that all operating costs, depreciation, and taxes would come out, and then anything left would be profit. Well, that seemed attractive to him and it worked out quite well, because we were able to increase the yield of peaches a few hundred tons the first year. Almost everybody was losing money at that time, but my thought was that I could improve on the management of the orchard. The big problem was that it cost more to produce peaches than you could get for the sale of the product.

We got the production up by using fertilizer. We also treated with zinc, a new thing recommended by Dr. Chandler out of Berkeley. It was just being introduced and I happened to have the excellent guidance of a very astute person on whose ranch much of the research had been done. Well, all the trees were starving for nitrogen and for zinc. With that, plus a little change in pruning and cultural practices, we came through with a 50 percent increase in tonnage in one year.

Then the market began to improve, so we purchased the property. Or, I should say, we signed a contract for the purchase in 1938, on January 17. And then it started to rain. We had the heaviest rainfall in 75 years. The farm had been leveled before the peaches were planted, primarily by horses. And when the water came there was no surface drainage. The soils in this area are relatively shallow, and this was in the days before we had high-powered tractors and subsoilers. In March I went up on the hillside in the back of the ranch to look over the fields, and I could see the trees were beginning to suffer from wet feet. I was standing there with a friend one morning, and he looked over the top of the orchard and said, "What's the matter with those trees?" Getting an aerial view of it was entirely different from looking from the bottom up. We went out in the field and did some further examination, and then I knew what was ahead of us. From that point on it wasn't too rosy a situation.

Actually I thought I was doing quite well until one day Cornelia at lunch said, "What's the matter? You look bothered." So I told her the story of what was impending. Well, Bob was about two years old, and Nancy was just three months old. And Cornelia said, "Well, Nancy did the cutest thing this morning!" I've never forgotten that part of it!

The trees started dying in March and every day there were more trees that died, and this continued from then until fall. Only the trees on the better-drained portions of the orchard survived. We had a ten-year contract with Del Monte at $25 a ton, but instead of having 1,150 tons, we came up with 250 tons. In the meantime we had dug seven-inch drainage wells by hand. Four of them to drain the water off through perforated casings. We dug trenches by hand, four feet deep, and filled them with gravel. As the trees died, spray outfits were stuck in the fields, 200 gallon tanks with four horses — the mire dragged us down. It was a disaster, particularly for a young farmer just getting started on his own.

At the end of the year our net worth was a negative $10,000. We owed the butcher, the baker, and the candlestick maker. We appraised the situation and decided that, given time, we could work it out. We wrote letters to all our creditors and told them the situation meant one of two things. If I went through bankruptcy, I could probably pay off 40 cents on the dollar. If they waited, they could get 100 cents on the dollar. Well, nobody pressed us for the money, and we were able to finance the next year by wood sales.

To get rid of the trees (this was before bulldozers were available), we ran ads in the paper and people came out and sawed the trees down and hauled the wood out. We provided a tractor to pull up the stumps, and they sawed up the stumps. We cleared the ground like that. I drove the tractor all winter. We ran the whole ranch ourselves until late in the spring when we were able to get some promise of financing. When they were looking for jobs, I told the pruners for the surviving trees that there wouldn't be any money until they got finished. The day they finished I had problems getting the money together that I had promised them. Instead of paying them at noon, I didn't pay them until three o'clock — they did get paid, though!

I've always been indebted to this community for the strong support they gave me under those conditions. It must also have been true that my creditors had some sense of what I might be able to do; anyway, they seemed to have faith in me, or were at least willing to take a risk. Finally we were able to pay off all our debts. We have done business with many of those people ever since.

We planted additional trees several years later. I releveled the ground with horses and one tractor and provided surface drainage. In 1946 we planned on starting in the grape business — we had grown some nursery stock and were ready to plant that. But Pacific Fruit Products, who had developed the fruit cocktail business, wanted to go into the frozen berry business. They conceived the idea of putting fruit in square cans, to save freezer space, and freezing them by immersing them in $-20^o$ alcohol. They wanted boysenberries because they were a high-flavored product. They approached us with the proposition of planting berries instead of grapes. So we sold the grape cuttings and went into the berry business. We bought plants from Knotts Berry Farm and a nursery in Watsonville. We got into a late planting season and made many of the mistakes that people do getting into a new business.

Boysenberries had been grown in Stanislaus County in the early 1920s in home gardens but not commercially. We planted 57 acres of boysenberries and wound up with something like 54 percent of the stand. In 1946 boysenberries were 22 cents a pound, and growers had about 1,600 acres in California. There was a heavy planting that year in this area and around Fresno. In 1947, when the inventories of freezers emptied during the war years had built up again, there was a slowdown in the market. Freezers carried over heavy inventories of all commodities. In 1947 we had a backup, and berries that year sold for 6 cents a pound. We had gone into the business thinking that the probable cost would be about 16 cents a pound, and that's where it was. Fortunately, we were diversified at that time. We had peaches and had started in the dairy business in 1942, and had some field crops, and then credit from the cannery carried us through that hump.

In the meantime my brother-in-law, who had worked for the Los Angeles County Agricultural Commission office, came up and bought a ranch adjoining us — and we were in partnership in the berry operation for 20 years. He got baptized by fire in the first year! Then he went into real estate, and when Bob got out of college I went into partnership with him.

We started out in the dairy business in a very small way. We bought 20 cows to start with. In fact, I milked those by hand for about six weeks before I could buy a used milking machine. This was a very strong Jersey area, because we had some good breeders here, and butterfat was the main market then. That operation gradually grew. During World War II one of the men who had been working for us saved a little money and bought some more cows, and we went into a partnership with him for a time. He went on his own, and another neighbor had a herd and was short of feed, so we made a combination with him for a while. Then about that time my niece's husband got out of the Navy and wanted to get into farming, so he took over the dairy partnership. After several years he decided to get into the insurance business, so that's when Bob took over.

When we started, 20 cows were considered a string — based on the number of cows that one man could milk. Then it was 40, and then it was 80, then 100, and now it's 200. Two hundred cows are in the same category that 20 cows were 40 years ago, a small herd. The production per cow is at least three times what it was then. Any time a cow then made 400 or 500 pounds of butterfat in a year she got her name in the paper. Now if she does that on her first lactation she goes to the butcher.

Our berry operation at one time was up to 132 acres. We farmed four patches. Now it's down to 40 acres. This was brought about by uncertainty of the market. We were competing with other colored pie fruits — for example, red sour pitted cherries — in the eastern United States, Oregon, and Washington — and blueberries. In 1954 there were 5,400 acres of boysenberries in California; now there are less than 500 acres. We weren't able to compete with mechanized-harvest commodities in other areas, and so the price dropped on down. When we had 5,400 acres in California, most of that was in 10-, 15-, 20-acre fields, just small family farm operations.

One reason we stayed in was simply because we were diversified. We were able to take one or two years of loss with the idea that eventually the fruit cycle would break. Sometimes in the East they have cherry frost failures, and then the market for berries bounces back up. On the average, we were able to come up with a profit. Boysenberries are very tied up with what happens to another commodity. We watch the red cherry and blueberry market as much as we watch the boysenberry market, because that's what it's hinged on. At one time, Oregon grew as many boysens by acre as we did, but we grow almost twice as many boysenberries by tons. Their yields are not as high. On evergreen blackberries, of course, they get phenomenal yields and outproduce us here.

Boysenberries have traditionally always gone to processing in one way or another, not to the fresh market. They can't compete with strawberries on the fresh market because their shelf life is very short. If they're frozen immediately, they'll keep. I don't know how long, for sure, but I had a No. 10 can of boysenberries that was frozen in 1946. I took it up to the Davis laboratory about 20 years later and it was still in good condition.

Then, of course, there is the risk factor. If berries need to be harvested today and you don't harvest them, you might harvest them tomorrow. But if the temperature gets to 95 and 100 degrees (which it does sometimes), then the crop is gone in three days.

We didn't think, with a small acreage of berries, that labor organizing would ever have been particularly focalized on this industry. But in 1960 we were harvesting 130 acres of berries, and right at the height of the June picking we had a labor strike. It was the beginning of Cesar Chavez's activities, and was an exercise in organization to find out how it's done. But it was most disruptive. We had probably 600 domestic pickers in the fields, and, as I recall, around 120 to 150 Mexican nationals, whom we were using only in the peak of the season. Well, when a strike occurs, the first thing the labor department does is to say you can't use the nationals. So we were shut down. We had hearings in Modesto which took about three days to set up. Bob and I were in hearings from 9 o'clock in the morning to 9 o'clock at night. Cornelia and Gwen were dispatching the trucks. We were picking berries all over, in four different areas, and the girls were managing the whole thing. The final decision was that it was not a bona fide strike, and we were able to finish the season by using Mexican nationals.

Then in 1963 another unionizing effort was made. That also failed, because there was no support in the field at all. The Mexican nationals were not interested because they were only interested in working. Some of the domestic people stayed away because they were intimidated. Over 50 percent of our crew was school youngsters, and the parents didn't want to get them involved in picket lines, so we lost pickers then. But many of the adults just didn't want to have anything to do with the unions. I think probably only about 5 percent of the people who came to pick berries had the express purpose to start a strike.

CORNELIA SVENSON SCHOPPE, AGE 70

Both my parents came from Sweden. I was born in San Francisco in 1908, and attended school there until seventh grade. My dad had been a contractor, but after World War I building wasn't profitable, so he decided to move to the valley in 1920. My folks had bought a 40-acre parcel southwest of Gallup and Dad thought he would improve the place. I graduated from the old grammar school, which is no more, and attended Gallup High School. In 1925 my dad was killed in a runaway horse and wagon accident. My ambitions to attend college were cut off, so I went to work after graduating in 1926.

I first became a reporter for the Gallup *Journal* for about a year. I enjoyed the work, but it was rugged, tracking news stories all over town, and since I did not have a car it was difficult. So I obtained a position as a bookkeeper for a hardware store. After about two years that store went out of business, and I went to Stockton and was a bookkeeper there for an automobile agency. It was the Depression by then, and this agency also closed. I came home and was out of work for about a year. In the meantime I was helping my mother supplement earnings on the farm, which were very poor. In 1934 I became bookkeeper for the Robert Jenkins ranches, directly west of where we are now. That's how I met Walter, who was ranch foreman, and the following spring we were married.

We moved directly onto this place, which we rented. In 1936 Bob was born, and things looked fairly bright that fall, so we decided to buy the place. The next year our daughter was born, and we were just getting things through to buy the ranch. It turned out to be a year of terrific rainfalls, which flooded out practically the whole orchard. There is a layer of hardpan under the soil, and water just stood there and rotted the tree roots. We almost lost the place, but the man from whom we had bought it was very considerate and we were able to stay on. My husband is very persistent. He wouldn't give up. He said, "We're going to stay." So we pulled up the trees and planted other crops.

Over the years my chief contribution to the farm, other than keeping the household running smoothly, has been the bookkeeping. That has been quite a complicated endeavor. It is very detailed, because we work on a segregated table basis in which we segregate payroll between the different crops, between the peaches and the almonds, and the dairy and the berries. It becomes very involved. At one time we had 130 acres of berries. One patch was located on the other side of Highway 99, about eight miles west; two patches were ten miles in the opposite direction; and one patch was two miles from headquarters. So we were very spread out and it took plenty of running around to keep up with the payrolls. We tried to keep meticulous records on each individual picker. Each patch had a "head checker" who took care of the payroll, and then I took care of the final clearing of all the payrolls.

Bookkeeping time varies a great deal, but I'm at the desk at least eight hours a day during berry harvest. It has become much more complicated over the years, with government reports and all the details that we have to keep now. We have more records to keep each year. The past two years we've had state unemployment insurance, and this year we've had federal unemployment insurance. I'm not paid for being bookkeeper. I just enjoy the net income for the family. My doing it, though, has saved the cost of having that service provided by some outside person.

Both Bob and Nancy have said, "I can't remember ever feeling that I wasn't going to college." That was just a natural part of their schooling, to go right on to the university after they finished high school. Nancy majored in English and got her teaching credential. In 1963 she was a Rotary Foundation scholar and went to New Zealand for a year. Later she worked as a travel tour executive. She married in 1969. Her husband is a plant pathologist, and they have lived all over the world — Thailand, Korea, and Bolivia.

Both of our youngsters had good minds, and they profited by experiences on the farm, and particularly perhaps by the fact that they didn't grow up running with a crowd. They're individualistic; they've grown to be very self-sufficient, and yet they are also very congenial with other people. We were moderate parents. We tried to be reasonable with them. We are quite similar today in outlooks, even though we're different generations. In fact, Bob has taken over many of his dad's little idiosyncrasies — I see it more and more as time goes on. Bob's relationship with his father is really quite close. A lot of family continuity in farming depends on individuals. The father-son relationship has a great deal to do with it.

We consider our home community Kinsey and our business community Gallup. Kinsey is a small town. The population was only about 500 when we first came here, but there were four or five churches. Many people in the area were attracted to this town because the Friends Church settled here many years ago — the

Society of Friends, they call themselves Quakers. A very fine group of people. Through the years Kinsey has become more or less a bedroom community. People who live out here like the small-town environment and yet their jobs take them other places. The growth has been mostly residential. When we first came there was one grocery store, two or three service stations, and an old bank building. Now there are four or five service stations, a meat locker plant, even a new veterinary service. The old bank building is used as housing for farm laborers (and it's a despicable old building). There must be 15 or more families who live there.

Once I could go into Kinsey and know practically everyone I met. Now I can go into the post office and hardware store and grocery store and not meet one soul that I know. The population must be around 5,000. The school enrollment is up. When Bob graduated from high school there were 13 in the graduating class — hardly enough to make a chorus!

There have been times that I had to sacrifice some desires for household things in favor of equipment purchases. I can remember one year on Christmas Day, Walter led me by the hand outside and showed me a beautiful disc with a red bow on it. He said, "That's your Christmas present!" And I said, "That's really nice. I'm sorry I can't put it in the living room!" That has been one of the big jokes of our married life. But we have never been without, we have always had plenty. We have had a good life.

## BOB SCHOPPE, AGE 43

I was born in 1936. When I was three days old I moved into the house where I am living now with my family, and lived there as a child until we moved into my parents' new house in 1953. I went through local schools and then to the university for four years. From there I went two years into the service. Afterward I went back to the university in the ag business management program, and got my Master's in 1961. In the meantime, I wound up marrying the girl whose dad managed the ranch that joins this one on the back — the home-town girl. Gwen and I moved back into the old house on the farm after I got my Master's, and we've been there ever since. We have three boys and one girl.

At the time I went into the service, we had about 130 acres of berries. We were using large crews — we had almost 1,000 people on the payroll one day, I remember, and we picked almost 100 tons of berries. I could see that if I was seriously interested in farming, it would be useful for me to learn the leadership techniques that they taught in ROTC and in the Army. My first choice when I went into the Army was the Quartermaster Corps, because I was a pomology major and I was quite interested in the transportation and the handling of perishable commodities — what happens to the fruit when it leaves the ranch, the distribution system, and so on. But I was a reserve officer and was going to be in for just two years. In 1957 things were kind of quiet, and only the career people were getting those assignments. My second choice was infantry, and that's what I wound up with. I was in the basic training outfit, so I did work with large groups.

I had the idea that I could transfer that training in the military over to my own farming operation, and it was certainly applicable. It's just what we'd been doing on the farm. It was easy for me in the service, because the people that I worked with were exactly the same type of people that we'd worked with here. Many of the NCOs who came out of World War II were just farm boys who got drafted. My folks went down to Texas when I was there and met some of the fellows I was working with, and for them it was just like coming home.

When I left the service, my uncle and my dad were in partnership. It was Schoppe-Clark Berry Farms at that time. We picked up another 40 acres of berries so I went in, and that was set up as a partnership between Dad and myself. My uncle decided to retire in the early 1960s. But for a while there was a four-cornered setup. Dad and I had a partnership, Dad ran part of the farm on his own, the dairy was in partnership with another cousin, and there was the berry ownership with my uncle. Dad had a lot of irons in the fire!

The ranch at that time was about 265 acres. We were also leasing 100 acres. When my uncle retired, Dad and I were partners in the berries, but the folks retained interest in the trees, and in the open ground and the dairy operation. In 1965 my cousin, who was in partnership with Dad in the dairy, decided that he wanted to get out — butterfat was 9 cents a pound then, and the business wasn't very lucrative. So I bought out his share, and we went into a full-fledged partnership on the whole ranch. Since then, we've cut back on our berry operation because of the risks, because it is so labor-intensive and extremely expensive. We've increased the dairy operation from 72 cows to about 200 cows. We're farming 300 acres that either the folks or Gwen and I own, plus roughly another 100 acres that we lease. The current makeup is 40 acres of berries, 40 acres of trees (almonds and peaches), and the balance roughly 20 acres in the dairy site, plus or minus 300 acres of field crops which are double-cropped corn and oats, or beans and oats. We've been phasing out of the trees and berries and into the open land crops and the dairy.

Actually, we've had some fantastic years in berries in the last five years. But before that the losses that we took were pretty rough. Berries take about 700 hours of hand labor per acre, for wrapping and pruning and the harvest itself. We see ahead of us increases in the minimum wage, which means that the financial risk is going to be even more. We want the balance wheel of the dairy operation to be large enough to take care of some of the years of the berries. We have had as high as five-figure losses off the 40 acres. It takes a pretty stable operation to absorb that kind of pounding.

The only berries that we grow are boysenberries. We had 12 acres of ollallies once. Some people confuse the ollallieberry with the boysenberry. A lolley is basically a blackberry. It's a long, narrow, cylindrical, shiny berry, high in acid and low in flavor. The boysenberry is much more succulent, high in sugar and better flavor for freezing.

We've gone in different marketing directions over the years. We started with canning berries, then Pacific Fruits from Modesto froze them. Our berries have gone to juice outfits, wineries. We've never sold to fresh markets. Fresh boysenberries are not conducive to shipment because they're high in sugar, and the molds come into them immediately. It's not the type of thing that you can put out tonight and if it doesn't look good you can still sell it in the morning. If it doesn't go tonight, it's in the garbage. Boysens are much more perishable than strawberries.

When I got out of the service, we didn't have much problem filling out our crews. We had the bracero program to fill in any gaps in our local labor supply. We used braceros till the last year that they were available — 1965, I think. About then we pushed out 40 acres of berries and planted it to almonds. As the other plantings phased out we just didn't replace them. Our last 40-acre field was planted in 1965 and that's where we've stayed. We're not sure we're going to plant anymore. The life expectancy of berries is anywhere from 5 to 15 years. We keep pruning that field on a year-to-year basis and it looks as good now as it did five years ago.

It's an expensive crop to establish. You have probably $1,500 to $1,800 per acre in it before you pick your first berry, and a year and a half of waiting. Not quite as bad as trees, but an awful lot of hand labor. A fellow who's been with us since 1944 looks after the berry operation and worries about it. He's an excellent manager. The day that he decides to retire may be the day we go out of the berry business.

When things got hot in 1967 and 1968 during the Viet Nam thing, it was difficult to get our berries picked, and we had to rely on school youngsters. It took the cooperation of the schools and the farm labor department. Then the whole labor thing became quite politicized because the union movement with Chavez came on. On top of labor problems, right after Viet Nam there was a mild recession right at the end of the Johnson regime. Food costs became a concern. People began getting out of berries. We're a very small industry now. There are only 350 acres of boysenberries left in the entire state.

When you get into a deal where you could lose $30,000 to $35,000 on 40 acres, we're such small potatoes that we can't afford it. The prices on berries are extremely cyclical. Since 1970 we have gone from 17 cents per pound to 35 cents, back down to 17 cents, and then up again to 20 cents. Year before last it was 42 cents, and we thought that was completely out of sight. Last year it was 65 cents — so berries have been extremely profitable the last two years. The funny part is that it's attributable to the overnight popularity of flavored yogurt. Boysenberry-flavored yogurt just took off, particularly in the Pittsburgh and Philadelphia areas. All of a sudden they couldn't get their hands on enough boysenberries. We're like the hula hoop, or pet rocks! You take it while you can and hope it'll last.

The schedule in berry growing is like this. To get started, you clean up the ground in the summertime. Then you fumigate in September, because berries are quite susceptible to nematode. Fumigation also kills the noxious weeds, Johnson grass, and Bermuda. In late January to mid-February you plant. Then the following August you put in your endposts and trellises, string your wires, and do the initial wrapping. Through that following winter you rewrap, getting as much foliage up on the wires as possible. Your first crop would come off the following June, roughly 18 months from planting time. After that, it's pick in June, prune and wrap in July, November through January the winter wrap, then spraying and spring care, and the harvest in June again.

We get up to about 200 workers now in the harvest; it takes about 80 for the pruning and wrapping. Our winter crew is normally five to ten. For the peak times we're fortunate to hit a gap in the local labor movement. We catch the families right after peach thinning and just before apricot harvest on the west side. The word gets out and we've had full crews on opening days for the last five years.

We pay on the piece rate, so much a crate, 14 1/2 pounds. We watch the bellwethers of the picking crews and adjust the rate up towards the end of the season. Last year we started at $1.25 and ended at $3.75 per crate. We try to keep the earnings of the good pickers on a reasonable level, as the picking gets more sparse. The first berries, on top of the vine and out in the open, are easy to pick. In the latter part of the season, the shadier berries underneath are harder to find and more work.

It is a real skill to pick berries. The range in earnings goes from probably a dollar an hour to six or seven dollars, on a sustained basis.

We have family groups out there, sometimes six or eight in one group. Our labor force is more of Mexican background now. We still have white school children who come out, but not nearly as much. The school years are winding up later each year, and sometimes youngsters get here after our crews are pretty well filled. In July we do use youngsters in the pruning and wrapping, maybe a third of the crew.

Some union organizing has happened in this area. We were the first ones to have Mr. Smith from the Agricultural Workers Organizing Committee in the early 1960s. They worked us over pretty good. We were the largest berry grower and they thought we'd be the most vulnerable. But there was almost no support among the workers.

One of the first things that we noticed was station wagons parked by the field. We normally start picking at 6 a.m. and I noticed two or three station wagons with signs printed up, all set to go. About nine o'clock there were people spotted in each crew that we had. They would start talking and trying to move people out of the fields. Some of it was physical, some of it was just persuasion. Our instructions from our attorney were to let the workers go out of the field onto the road and let it go from there. That's the way we handled it. It was obviously a planned tactic, nothing spontaneous about it at all. Most of our workers didn't respond particularly. We were picking about 2,500 crates a day. The first day of the strike it dropped down to about 1,500, and then after that we were about the same. We didn't lose fruit any of the four years.

I don't think the labor situation is such a philosophical issue as it's painted to be, that there's got to be a conflict between growers and farmworkers. The workers are doing the same thing that I'm doing, they're looking to make a living, to make a dollar. If they don't like working in berries, they're not going to work here. I'm not married to boysenberries, either, and I'll pull them out if we get into a bad situation. The risk of the business itself is such that we can't afford the other uncertainty. If you have a contract, you may get a slow-down at harvest time. To me it seems you're just subject to blackmail at any time. We have the dairy and might be subject to the same blackmail, but I can step in there and do the milking. But I can't step in and pick 40 acres of boysenberries.

I don't see California ever getting back to the 5,000 acres of berries that we had in the state at one time. I'll never have more berries than I can afford to walk away from. If I had 500 cows milking, I might have 80 acres of berries, but so long as I have 200 cows, I'll have 40 acres and that's it. If you're highly levered financially, you don't want to be in the berry business. There used to be a lot of 10- or 15-acre growers, sometimes with part-time jobs. The risk now is such that it might wipe out all the earnings from the part-time job. You could lose $8,000 to $10,000 on 10 acres of berries easily.

We've invested money in mechanization projects through the Bush Berry Advisory Board. The problem with the boysenberry is that it is difficult to separate the berry at the cap stem. A mechanical picker tends to get the little two-inch fruit spur with it. Then when it rolls along the conveyor belt it damages the adjacent berries. As soon as they start bleeding, mold takes over, and there is a sticky, gooey mess along with the leaves that have also been shaken out of the vines. When we froze the berries in the field, putting the berries through a freon bath, we had a beautiful product. They came off the end of the conveyor before they had time to congeal into a mass. The berries were frozen to $-73^\circ$ C within a tenth of a second. The thrip exploded out of the berries, the dirt didn't adhere. It was very clean. But the expense just was prohibitive. Mechanical harvesting might be feasible now for a puree-type product strictly for preserves. But an IQF (individually quick frozen) berry for the pie trade isn't going to be feasible until we have field freezing, with a unit light enough to take both the machine picker and the freezer down the road.

We sell plants to a fellow who runs a U-pick operation on the other side of Claypool. People come down on the bus from San Francisco, pick berries out in the heat, and pay almost more than they can buy them already picked at roadside. Just for the experience, I guess, of picking the darn thing! On a salvage basis, on the last pick, we have let people go in and pick their own berries. But our experience has been that a novice picker does more damage to the vines for the succeeding year's crop than we could compensate for by the sale of fruit. We train the berry plant upright on the trellis, but the cane which bears fruit for the following year comes out of the base of the plant and lies along the ground. If novices get in and thrash around, the new growth is really damaged.

Labor isn't as much a risk factor in peaches, although it is a factor. To my thinking, peaches just don't have a market. They're in an extremely competitive situation. When they sold canned peaches retail at three for a dollar as loss-leaders, we thought that we could sell 25 to 30 million cases, but as it's worked out since, the value of fruit in the can doesn't match that of the label, or the cost of the sugar and everything else. Seven halves of peaches in a small can is not a value at today's prices. The demand for canned peaches has fallen way off. This last year they're going to market maybe 21 million cases, and yet the supply has been 25 to 26 million. So there's severe price cutting.

Part of the reason for diminished demand is the costs that canners have built into their operation. Energy costs, production costs, the costs of a union contract that was signed three years ago. Compare peaches with boysenberries: When we sold an IQF berry this last year, I think the market opened at 90 cents. Of that, 65 percent was raw product payment. Those berries were being enclosed in a poly bag to go directly to the housewife. On the other hand, if you get 60 cents a can for peaches, there's only a nickel for the fruit in there. Everybody else has more control over the price than the poor Yahoo putting peaches into the cannery at the beginning end.

Yet growers own this particular canning facility. We're up against almost a philosophical issue. The labor strikes are not so much over money matters as over seniority (there are two seniority lists within the cannery). They had a big revision of the seniority list. So when you take fruit in on a Saturday when they're not running the normal can lines, the people who have worked on the grading belts can come out and bump the forklift drivers. So you've got Saturday forklift drivers who have never driven a forklift before. To me it's a tremendous amount of inefficiency. I don't see that the canning peach industry is going to come back.

At one time 10 percent of the peaches grown in the world went across one slab up here in Echo where we used to deliver. Now you drive through that country and it's acre after acre of almonds. The only thing that will save the peach industry, I think, is mechanization. If everybody who works in a peach orchard is supposed to make $15,000 a year or whatever the goals of the social planners are, there's no way that anybody is going to buy the product. That will be the straw that breaks the camel's back as far as the product's value in the can. The regulations, the expense of handling people, are incredible. We just tallied up the other day and our payroll costs, in addition to federal unemployment, are 25.1 percent. Then we have health insurance and vacations for our year-round fellows, which add another 5 or 6 percent. You just have to have the productivity to cover those extra costs.

I thought about this yesterday during a meeting and worked it out. We've developed over a million pounds of milk per dairy employee. With that type of productivity you can afford to pay labor and all the extras, but if you're talking about peaches picked by hand, then the profits just aren't there.

Our dairy employees are quite different from our field workers. We're still in the same dairy barn after 12 years. Almost everybody else has updated their barn — we're going to as soon as we can. We've been Grade A since 1953. When I took over the dairy we were milking 72 Jerseys. We expanded the herd to about 125 when we made the decision to convert to Holsteins.

They produce more milk and less cream. The tendency at that time was away from butterfat and cholesterol. We're producing meat as well as milk, and Holsteins are more efficient beef converters too.

We started with one milker and one fellow doing the feeding and calf feeding and the whole works. Since then we've increased the operation, so now one fellow supervises the dairy, there are two full-time milkers, and one fellow milks three days a week. Another fellow feeds six days a week, and a relief man fills in and feeds one day a week. They're all living in the area. Three of them live on the ranch.

I still do as much physical labor as I can. The main part of my time I do straight supervising, mainly trouble-shooting and filling in. Sometimes I wind up loading trucks and moving boxes. I still do some field and tractor work. Usually I try to have things started in the morning when the regular fellows come in, and then they just pick it up and go. Then I usually go at night again, unless I have meetings. With irrigating, the fellows watch the water on this place, and my boys and I take the outlying places that we lease. We've always hauled our own hay too. Gwen drives the truck for me, and the crew loads and I stack. We're going to get away from that, though. We built a hay shelter with the proceeds of a good berry crop, and we built up our dairy operation, so we'll be stacking hay mechanically.

Almost everybody who is staying in the dairy business is expanding. I hardly know of a dairy operation the same size as it was ten years ago. Expansion is a necessity just to stay in business. I just looked at the time tag for Steve Vanders. He'll make close to $280 this week, not what you think of as poor, dumb, unskilled farm labor. But to pay him what he's worth, my figures show I have to get roughly a million pounds of milk to cover his job. As the economics change, that figure may very well move up to a million and a half pounds.

I'm going to meet a fellow at 3 o'clock. I want to buy a six-row corn planter. It will plant half again as fast as the one that we currently have, plus it's got a cross conveyor for fertilizer, which will shorten up our downtime. We'll be spending about $9,000 in order to stretch the labor that is available — or my own time if I have to do it.

In this area some very broad changes have taken place since I've been farming. There's been a lot of consolidation. For example: In addition to what I think of as our home place, places that formerly provided livelihoods for four families have been consolidated. Where there were five units there is now one. Where have those other families gone? Well, one woman that we lease 35 acres from is a part-time teacher for handicapped children. That place was formerly a full-time dairy. We took down the barns and cleared the ground for farming. Another fellow we rent from is a cattle dealer now. Another 20-acre piece is owned by the principal in the local school. Her husband formerly farmed that ranch in addition to some other custom work that he did. Then the 40-acre berry patch on the corner was at one time a 40-acre farm of alfalfa, sweet potatoes, and watermelons. Those small acreages, instead of being sustaining operations for somebody, have become supplementary income as the owners rent them out to someone else to farm.

There was a lot of fallout from farming in the late 1950s and early 1960s. Now the land for sale around us is developing into homes-in-the-country and pony-pasture type things. There are seven houses on 20 acres not far from here — and three acres of horses. We're getting more and more of that on either side of us. I counted up one time and we had 32 neighbors. One is a pilot for TWA, another for Flying Tiger Airlines. One is a teacher. And so on. We're becoming urbanized, just like everyone else.

But the fellows who have sold their ranches just can't stand to retire completely. Some excellent people have helped me work ground. The difference in equipment was really brought home to me this last winter. A fellow who had been planting grain for me started one morning at 6:30. We had a 13 foot, 6-inch drill, which takes 1,800 pounds each time we fill the hoppers. When he went home at 5:30 he had planted over 75 acres of grain. He was just shaking his head, because he said that if he planted 20 acres a day in the early 1950s he'd had a big day. So now it's easy, but it's capital intensive. It's a $5,000 drill and a $10,000 tractor, whereas in 1953 a 10-foot drill cost $750 and a tractor was probably $3,000.

Consolidation is a fact of life. We see it in everything else. There are just three major automobile manufacturers. There is one major transportation union for the entire nation. There is a federation of government employees, which is going to wind up controlling all governments — local, county, state, nationwide. Then people say only the affluent can afford to farm. A comment was made by a fellow at a peach meeting recently: A farmer is a guy whose accountant tells him he's a millionaire but whose wife doesn't think that she can buy a rug! And that happens to be just exactly the situation that we're in. We've found that if you get behind in keeping up with your equipment, you're just dead in the water. Either you farm and stay up with the changes, and spend your money there, or you don't compete.

In 1965 Gwen and I started buying ground. We bought 20 acres on our own, then we bought 40 acres jointly with my parents. Then we made the decision that any further capital investment would be done in the dairy rather than in land. About six years ago Dad said, "It's time for me to retire, so you're going to do the buying and the selling. I don't want to know who we owe, how much we owe, who owes us, or how much. You take care of the business end of it, and I'm going to drive the tractor, irrigate, I'm going to farm." I thought I had the easy part, but he really got me! That's the way it's been for the last six years. Since then Dad's become extremely interested in Rotary Club activities, and an antique equipment club, and I've been farming. It's the way that I would like to do it with my family as well.

The future of the family farm in California is more hopeful now, I think, than it was at the end of the 1950s. The premium for any hired labor is going to be such that the value of the individual's labor is going to increase as well. If he can be fairly self-sufficient (which is the attractive thing about almonds) and get some part-time family help or fill-in help when needed, then one family can manage a 120-acre almond farm, for example, efficiently. I think dairying will be good in the future. The fellow who used to milk for me until about four years ago came as an immigrant from the Azores. He was able to bring $5,000 or $6,000 from the old country. He milked for me for four years and he was extremely frugal -- and now he's milking his own 125 cows, and he's got his own feed and his own equipment, just in ten years. He doesn't make all the dances and the social events, but if you were going to put a figure on it, the guy is worth six figures at least. It takes work to farm successfully, but you can still do it.

I'm not particularly threatened by the thought of corporate farming, either personally or in general. Knott's Berry Farm is putting in 100 acres of berries up here, but it took them three weeks to get the contract to push trees out cleared by their board of directors in Anaheim. I don't think corporations can make it, particularly on a labor-intensive crop. We've seen other corporations really take a bath. If families can hang in there, they'll prove they're more efficient.

## GWEN HALVORSEN SCHOPPE, AGE 42

I was born in 1936 and raised on a cattle ranch just one road over. My father was the foreman, the owner lived elsewhere. That was my home until I went away to college and was married. I went to Pacific Lutheran in Washington and got an elementary teaching credential in 1958.

I met Bob about the second grade, because we went to the same schools. He was a year ahead. We were married in August after I graduated from college. He had ten months left in the service and we spent them in Texas. I taught girls' P.E. in an elementary school. I came back about two months ahead of Bob because I still had a few requirements for the California credential and I thought I would teach. Instead I worked in the public service office at the university while Bob finished his Master's.

Both of us always assumed we would be coming back to farm — at least we were going to give it a try. Our first son was born while we were at Davis, so I didn't teach for several years. Now we have four children, all within six years of each other. The youngest is a girl.

I didn't start teaching again until about eight years ago. Then I started substituting. I like teaching, and kindergarten is my love. But I enjoy working on the ranch, too. I drive trucks during hay season and when I am teaching I really miss that. I enjoy going with Bob to take a cow to a sale, or when he's branding calves.

Teaching brings home extra money. On a farm all your money is tied up in equity. Even if your books look good, if you're worth a lot, it doesn't give you money to spend on extra things. Teaching gives me the chance to buy a new couch if I want it. I'm bringing home, from another direction, what used to be the old farm wives' chicken or milk money — the cash for the extra things.

Bob and I decided a long time ago that our kids have a unique opportunity to work at home that kids in town don't have, and they will always have money available to them if they have earned it. When our boys become 13 they start working on the ranch. Until they're 13 they pretty much choose — "Would you like to do this? This is available. Do you need the money?" But when they get to be 13 they're pretty much on their own for spending money. They have bought their own clothes since they were 13. They've had savings accounts of their own since they were small. When they started working, half of their checks went into their savings accounts, and with the other half they could do as they pleased. I have one son who's tight as a bean, and he would keep out a dollar maybe, and put all the rest in savings. Others would spend more. When they graduate from the eighth grade they are given a checking account that they have to maintain. There has never been an allowance, or anything like that; they work for what they get.

My daughter is 12. There are not as many opportunities for work for her. She can pick berries and can clean up brush, and she washes down the dairy barn, which is an hour on Saturdays, but she doesn't earn much money. She's not strong enough to do some of the hard jobs that we have here, so the boys have determined among themselves that we ought to give her a couple more years before she earns her own spending money.

Our oldest son is at the university. If any of them farm, it will probably be him, but I hope that he doesn't feel trapped into it. I would never want him to farm under those conditions. But he has always liked the ranch and all phases of it. He likes the cows, and the truck work, and the field work, and the irrigation. Our second son doesn't like the dairy at all. I'd be surprised if he did stay on the ranch. He values his free time a little more than a farmer can. He might be happier in a 9-to-5 job where you get two weeks off, and you're not tied down like you are to the dairy or to a farm. Our third son is exceptionally interested in science and math, and very talented, and I expect him to find his niche somewhere in research or where he could use that ability.

I wouldn't doubt that Bob would be delighted to have them all stay farming. When we came back we said that we would give it ten years, and if we felt that it was not what we wanted to do, we'd still be young enough that we could change and do something else. Bob's folks really wanted him to come back, because he's an only son. As it turned out, we're very comfortable here. But we're quite conscious that people can be trapped into it, and we'd hate that for our sons.

The boys have had the opportunity to work with their dad and be with him a lot. Bob and the boys are close because they've had this working relationship all the time. The major drawback is that because we're diversified, we're busy all year round and there is not much leisure time. There's no time to take a two-week vacation. In fact, we've done it only once, and it was only five days that we left the ranch as a family and went somewhere. We can never work out the timing. We have a cabin in the mountains, and I always take the kids for two weeks in the summer. But Bob has not been up there for more than three days at a whack. He just can't stay away that long. I don't know if he actually can't, or that he just loves his work. I mean, he loves it!

He really works from dawn to dark. This is wintertime, and he's always out of the house by 7. He never comes back into the house except for maybe half an hour for lunch, and then he's off again. He seldom comes in for the night before 6 o'clock. When we were first married at times I resented the fact that he was gone so much. Our children were close in age, and I was pretty well tied to the house. There were times that I thought that a 9-to-5 job would be wonderful, with weekends free. Sundays could be a day when absolutely no one made a demand on you -- family time; you could go for a ride in the country. Well, I don't think his schedule has changed at all, but I think I have!

My last grandparent died when I was seven, and I never had a relationship growing up with my grandparents. I am so happy that my children have had the opportunity to live next door to their grandparents. It has been good for all of us. I wouldn't trade farm life for anything.

An irrigator wields a shovel as water spills into field furrows. Reprinted with permission from A Guidebook to California Agriculture, University of California Press.

# STRUGGLE TO SURVIVE
## The Webster Family

Bill Webster senior, and Ida: quiet people, not much prone to talk. Though they have been well acquainted with hard work and slim budgets, their home now is quietly elegant, tastefully furnished. Bill is tall, thin, wiry even at 74, laconic in speech, faintly distrustful of visitors who ask questions. White-haired Ida has a sense of style in her dress, but she too is sparing with words. Much remains unspoken.

BILL WEBSTER, SENIOR, AGE 74

I was born in 1904 in Ventura County. My dad came from Canada and my mother from Kentucky, way back in 18-something. My dad was a farmer. I lived on a farm all of my life. We were around Ventura County until 1918, and then we moved to the San Fernando Valley. My dad was there 12 years. He didn't have a very big farm down there, 20 acres. During the Depression things got bad, so he sold that, or traded it in on land up in Siskiyou County. He stayed one year on that, then he left and come down here to Stanislaus County. That was in 1931.

It was pretty barren around here. We leased land from the Wilson family out here. When the folks passed away, me and my brother bought this ranch in 1945. But all during the Depression we were renting. We were growing beans, just strictly beans — baby limas. All I ever grew was beans, and some alfalfa hay. Then after the kid took over, he got very diversified. It could be irrigated here by 1931, but there wasn't much of it done.

We leased about 250 acres of the Wilson ranch for 15 years or longer. We did most all the work ourselves, hired very few people. Me and my brother always worked together. My brother lived in Paisley in an old railroad house they used to have. I lived in the depot myself for a little while. There was no other place to live. Wilson finally had an empty house out on the ranch, and we moved out there.

We didn't know much difference if there was a Depression or not, because we hadn't had much before. Beans and grain was about all they had when we came here. No trees right around here. On the Leachman Ranch there used to be quite a bit of orchard, some grapes on it. My wife worked there cutting peaches. It was all big ranches, and they would rent out individual ranches to families like ours. The first year we was here, there wasn't many at it.

Some land was volunteer barley. One outfit tried to grow some cotton over here, but it was a failure. New ranchers would come in for a year and then leave. It was hard work. We hung on because instead of hiring it all done, we'd do it ourselves. My dad had a team of horses when he first came up north, but when we came down here he got rid of them, and we bought tractors. The Wilsons had to finance us for a few years and help us out. They had about 2,000 acres. They leased outside land, too. They had a cattle ranch up in the hills. It's hard to pick up land around now. That is a change since we started farming. Quite a change. Rent is getting so high that it is pretty hard.

One year of high school was as far as I went. I wanted to go to work. My folks got the police after me, trying to make me go to school, but I didn't care for school. I was 16 and the police said they couldn't do anything. I made a big mistake, I guess. But I lived through it all. I started out doing a man's work at 16. I worked on ranches. I worked on a big dairy ranch, drove a truck around, and worked in a gas station and a grocery store. But that just wasn't for me, I guess. I came along with my folks when they moved up here.

I got married in 1931. We had a little shack up back of Paisley that we lived in for a while. I worked with my father. Summertimes, we worked daylight till dark. In the wintertime you didn't have to do so much. But during the summer we sure worked. So much harder then than it is now, too — they got things pretty easy now. The old tractors we used to have to drive was mankillers, pulling them levers all day. Now they have hydraulic and all that, it just takes one finger.

The war brought prices up a little. In those early first years we just made expenses. Of course everything, water and all, was much cheaper then. If we had to have labor, it didn't amount to much. I worked out for two bits an hour a couple of years for some ranch that needed help for a few weeks.

Me and my brother debated a while before we bought this place. We leased the land for four or five years before we ever bought it. We should have bought it the first year we leased it, because we would have got it for half of what we paid. We hung on, and they doubled the price on us. We bought 340 acres, with a government loan, around 1945. Then we had cash to pay down a pretty good payment on it. We had 40 years to pay the loan off, but we paid it off a lot sooner — in less than ten years.

My brother was my partner until he passed away. Beans were my part of the ranch. I stuck with beans — they're dependable. They didn't have to be sold right when they were ready, like this other stuff. Bean market was high and low, but we always managed some way. We thought about getting bigger, but we thought it would just be a headache.

Me and my wife have just the one son. He was adopted in 1949. We had our application in a long time before that. We had three daughters, but they died as babies. When Bill was adopted, I had someone to carry on after me.

There wasn't much change in farming during the 1940s and 1950s. Prices picked up a little. Walnuts were pretty good for a number of years. We did pretty well in those five years after the war — built a nice home in 1950. My brother lived on the other end of the ranch, so I built down on this end.

When Bill was going to school he wanted to farm. I tried to change him — I thought it was a hard road the way everything was getting. All this spraying, and insecticides, and fertilizer, and bookkeeping — they got to do all that nowadays. It was getting so complicated, prices, machinery, and everything. Well, he couldn't have farmed at all if I hadn't already had the land and the machinery. I tried to discourage him — I thought he had better get an education and get some training. Farming was rough, and I knew it was. At times I am glad that he decided to stay with it. But the way the young generation farms, it is pretty expensive.

We rented our land out in the 1960s, for two or three years at a time. These tomato men only want the ground for two years. They were paying pretty good cash for the ground, and it was all their expense and work. So a couple of times we rented part of the ranch for tomatoes. Then we had alfalfa seed on this piece for six years. The last time I had hay here, I hired a baling crew.

Bill is in diversified crops. I never did go for that. It is a big expense for all that stuff. It is more risky the way he is doing it. You never know what you are going to get out of green tomatoes. When they are ready, they got to go, no matter what the market is.

Bill is starting with more at age 30 than I had. He doesn't have to work as hard as we did, either. He keeps a couple men the whole summer. He does a pretty good job of farming. He is careful. He could spend a little less than he is paying to buy something he probably don't need. I don't think he needs more acres, I don't think this big farming is so good. A big ranch over here went under two years ago. The nephews took over, and boy, they went bankrupt. They finally had to sell the place, all in big blocks.

We have less open land around. Along the foothills up there used to be all dryland grain farms until that canal went through. Now it is all orchard. There has been quite a change in the valley too. If some of them old-timers would come back and see their old places, they would probably roll over in their graves.

My wife's a good manager. The most important thing she's done for the farm is to be stingy. When we first started drying our walnuts, she helped out on the huller, picking out bad ones. But that was about all the farm work she ever did do. She started to do bookkeeping, but then I wouldn't go by them, she said, so she quit.

Farming is a good life. I don't have any regrets. We done pretty good. Now that I'm retired, we have a second home in the foothills and three, four times a year I go up there and spend time.

I don't know what will become of Bill as a farmer today, to tell the truth. But if anybody makes out, I think he will. He gets around, looking into everything. He finds out everything he can learn. He is on a share basis with us, instead of putting out cash rent for our ground. That helps him. If we make money, he makes money. But if he puts out cash rent and don't make a profit, he might just have to pay all his income for rent.

Some newcomers are coming into this area — Southern California people. And San Jose people came over here and bought land when their land went for subdivision. That's what's raised the price of our property. They got a big price for what they sold. So they come over here and pay a big price for land here. Then our taxes get higher.


IDA MACKEY WEBSTER, AGE 68

I was born in Endicott, Washington, on a wheat ranch. My grandfather was a wheat farmer. We moved to town when I was 8 years old, and came to California when I was 13. My dad had a chicken ranch in the San Fernando Valley. He was a carpenter too, in Burbank. When I graduated from high school I did some secretarial work, but at that time jobs weren't plentiful.

I met my husband near San Fernando. We were married down there and came up here. First we went to Bealeville for several months, and then moved here. It was very dry and barren and hot. He and his family were trying to get started in farming. There were supposed to be good opportunities here. First we lived in a little two-room place — the buildings are long gone. No shower, no toilet in the house. A little kitchenette in the bedroom. Then we moved upstairs over the railway depot and lived there for some time. And then we moved down on the Wilson ranch, where we were renting.

I was just keeping house, and it was lonesome for me. A relative of mine lived in Sunset, a seamstress, so I helped her. Seasonally I worked a couple of times, one time pitting apricots, another time on the sorting belt at the bean warehouse, to bring in a little extra money. It was difficult finding work during those years. Just that sort of work was all that was available.

We didn't have much social life or recreation, except family getting together. I didn't particularly want to stay here, but I am glad now that we did. At the time I thought I'd rather live anyplace but here. It was just too lonesome. I thought then that there was too much family, but it finally worked out all right. Paisley was tiny, but then Burbank wasn't much of a town either. I didn't miss it.

During the war we all had victory gardens. I had such a big one, I carted vegetables all over the countryside trying to get rid of them. But gardening isn't something that farm wives do much in this area, not any more.

We didn't encourage Bill to go into farming, ever. In fact, we tried very hard when he was in high school and afterward to encourage him to think about something else. I was afraid that he wanted to go into farming just because he thought so much of his Dad. But I thought his personality was such that he might be better suited for something else; he is very outgoing. But now I think he is doing very well, as long as he keeps all these other things going on the side to fill in the bare spots.

After we adopted him, things were going better for us. He has seen all the good years in farming, but he had never seen any of the struggle. Most of these young people now have never seen that, and might not cope too well if bad times came again. It takes certain very down-to-earth people to go into farming and stick with it. You have to learn to take the bad with the good, soon and early.

My husband and his brother always did all of the work they could. They never hired people unless they just had to. They never put off until tomorrow what they could do today. One of the problems with farming today is that more and more things are hired out. It makes a great deal of difference. Bill could do more than he does, in some ways. But of course they want to farm the big way. They expect to start with more than we did, and to end up with more.

Bill was always close with his father. He thinks there is nobody like his dad. We have been pleased that he came back to the area and started farming. It would have been very lonely if he hadn't. We have been very quiet people, private people. We don't get out a whole lot. We have traveled some, but pretty much we stay close to home.

Paisley, our little town close by, has grown immensely in the past few years. There are more farm workers in the neighborhood than there used to be, because of the different crops in recent years. Paisley has a number of new houses for Mexican people who do farm labor, but it is not a camp. They all own their own homes. The town is just about the only place for them, since it is less expensive. Paisley School is predominantly Mexican. You couldn't ask for nicer children and nicer families than the Mexican people who live here permanently. It is the transients who kind of spoil the schools.

The cultural things are not very handy out here — libraries and theaters and all — but we go into Modesto quite a bit. We do almost all our shopping there now, even groceries. We don't go much to Paisley.

    Bill and Shirley Webster live in a modest gray frame house behind a neighboring rancher's bigger, newer home. Their furnishings are sparse, but the rooms are clean and comfortable. A fire is burning in the brick fireplace on this gray winter day.

    Bill is stocky, with curling red hair. Emphatic in manner, his square hands saw the air as he makes verbal points. Underneath the positive exterior, however, one senses a strain of defensiveness. He is struggling to keep afloat in this competitive world, to hold on to the values his father taught him even as times and economics change. Shirley is sweet-faced, soft-voiced, slightly plump in her gray slacks and dark sweater. Whatever Bill wants is what she wants too.

## BILL WEBSTER, AGE 32

I am adopted. I never knew my real parents. As far as I am concerned, I was born and raised right on the ranch. My earliest recollections are all from here. I went through little Schirmer School just west of here, and on to Sunset High, and then to school at Cal Poly in San Luis Obispo, where I majored in agriculture and minored in English.

I served for three years in the Marine Corps with a tour of duty in Viet Nam. I was a machine gunner on a helicopter. When I got back from overseas, I met my wife at Camp Pendleton. We were married in November, I got out of the Marines in December, and we moved back to this area immediately. At first we lived in Sunset for a little while while I was working on my father's ranch. Then we moved to just behind Paisley for about a year, then we moved again down to Deer Crossing, a half mile from where we are now. We stayed there close to seven years. Then we moved into this place here, where we have been for two years. We have been married just over ten years. My son William Peter is nine, and my daughter Lisa Marie is six.

The military didn't do much for me. The only thing I picked up in the service — aside from memories (some pleasant and some not so pleasant) — was maybe the ability to get along with people a little better. Basically I was a farm boy thrown into a potpourri of people from all walks of life. In a situation like that, you are on your own, nobody fights your battles for you. You have to learn to get along with people. But that has helped substantially in my dealings with different businessmen, different companies, my own personnel on the ranch.

My parents did not encourage me to come back to farming, but it was all I ever wanted to do. They were willing to back me, and I think they are happy that I did choose their way of life, but I was never pushed into it. When I came back, I had made up my mind that I wanted to spend the rest of my life here.

Originally the ranch was the Webster Brothers, consisting of my father and his brother. In 1967 when my uncle died from a heart attack, I was in Viet Nam. My father bought out the partnership from my uncle's wife, so he became the sole owner. Today I call it Webster Farm. Back then, it was just Bill Webster Ranch, and my father leased the land, with the exception of one field, to one of our neighbors, a well-known big farmer. I went to work for my father for a year, until the lease was up on the other pieces of ground. I then leased from my father the upper ground, Field One. Shirley and I made our living on that — we didn't live well, but we made it. As my father retired, I took over more and more of the land. I don't know many young farmers in this area who farm on their own. Most of them farm with their fathers and are incorporated into the family business. But I am totally on my own.

I have always leased from my father and from my aunt. I also lease another piece on Lake Road. Presently I am farming 450 acres, all row crops. I grow a number of different things. This spring I will plant 200 acres of large lima beans, roughly 100 acres of green beans (freezer beans), or else dry baby lima beans. Then I will have 50 acres of green tomatoes for the fresh market, and 50 acres of alfalfa hay that will be third year. I do grow winter crops on occasion, but not this year. Last year I had 100 acres of winter peas. We have had wheat.

We had a walnut orchard but pulled it out four years ago. The orchard was 43 years old, production was dropping, and I didn't feel we could spend the money to mechanize the harvest — we were still picking by hand — and come out on it in the long run, due to the limited years that the orchard could remain in production. So I went ahead and pulled it and planted the ground to alfalfa.

In this area we are sitting on some very good ground. Some of it runs 100 percent on the Storie Soil Index. The only thing that limits us is the climate. The water is here, at least if you can afford it. During the drought I drilled a well that pumps close to five acre-feet of water. So I have a readily available supply of water, and by today's prices nominal. It costs me $7.11 a foot, direct energy cost, to pump (that is nothing for the amortization of the well itself). But water purchased through the district costs $14 an acre-foot, so I pump my own water for half of what it would cost to buy it — if you discount the $30,000 investment of drilling the well!

In farming you are either in a business or out of it. If you are a bean grower, you are a bean grower. Good or poor prices, you still plant, because of your equipment and machinery. It's hard to second-guess the price each year. The person who tries to get in a certain crop on any given year, hoping to hit the market, usually comes out second best. He might do it once or twice, but sooner or later he'll get hit. If you stay in a crop, you hit the highs and the lows both, but on average you'll come out. My business is built on the average. Basically, I plant pretty much the same every year, except for rotation. We do rotate crops in and out of different pieces of ground so we don't tire the ground.

My 50 acres of tomatoes are all hand-harvested. The risk factor is very high for the fresh market, you never know what will happen. A strike, rain damage, a freeze, a hundred different things can ruin the season. The tomatoes are marketed through a local company — I am under contract to grow a set number of acres. The contract is renewable every year; either of us can terminate it. They take everything I grow on those acres, which to my way of thinking is better than contracting tons or boxes.

My freezer beans are also on contract for a frozen food company. It is not automatic that you get a new contract every year; the company can tell a sloppy grower to get lost. On the other hand, if you have a crop failure, that doesn't necessarily mean they won't give you a contract the next year. They look at the averages too. They say, "This grower has a five-year average of 3,500 pounds to the acre, but he only got 1,500 this year. There must have been something wrong with the year."

My alfalfa hay is marketed through a cooperative. My dry beans — the large limas and the baby limas — are marketed independently. I store them; I have a warehouse full. The only thing that limits how long you can carry them is how big your pocketbook is. The price of dry beans is linked with some of the other commodities. Not so much corn, which is a grain, but with the Michigan pea bean which is grown extensively back in the Midwest. If they have a real big crop and an oversupply of the Michigan pea bean, then our lima prices go down. Generally, as a rule of thumb, you figure on $8 to $15 difference between baby limas and large limas. If the baby lima price goes down, the large lima price will too. Black-eyed beans are also linked very closely.

I looked into joining the Bean Association, as did my father, who grew beans for years and years. I didn't join it for the same reason he didn't. I don't like to lose control of my crop. I am kind of an independent cuss. If the market is $35 and I feel like selling, I want to be able to sell. If I sell for $35 and the price goes up $2, I will kick myself and say, "Why, you dummy." But if I sell for $35 and three days later the market goes down to $33, then I can say, "Boy, aren't you a smart cookie." The association handles a large volume and they sell a certain amount every month. Whereas as an independent I pick the time to sell, and there is nobody to blame but myself. The gamble is worth taking, for me.

My father never grew a large lima bean, or any peas, and only grew green beans once in his life. He never grew any tomatoes. It makes sense to diversify if your equipment can be used in more than one crop and you have the type of ground that will support three or four different things. You are better off not to put all your eggs in one basket.

I would like to have a walnut orchard, but to plant an orchard on ground you don't own is not good business, unless you can get a long-term lease, 30 years or more. I wouldn't be afraid to plant trees on the land that I lease from my father, but the land that I lease on Lake Road is only on a year-to-year basis. My aunt's land is mine to lease as long as I want it. She is 68 years old. Upon her death I have a two-year lease, and then it goes to her kids. There are two daughters: one lives in Modesto and one lives in Canada. If it comes up for sale, I would like to be in a position to buy. I don't own any land myself now.

When I started farming there were only two ways to go about it: One was to buy a little piece of ground and try to make a living on it. The other way was to start off small and then expand by leasing. The first year I farmed, I had 50 acres.

Now I have 450, in a span of ten years. Growing from 50 to 100 acres was very, very hard; going from 100 to 200 was very, very hard. But going from 350 to 450 is not so hard, because you can add 100 acres more easily when it is not such a large percentage of your total operation. My idea was to expand and get to the optimum size I wanted to be. I risk everything I own, every year, to do that.

I have a dream: Someday I would like to own 640 acres — a section of ground — farm it, and have the money and everything that goes along with it. That is my goal. I think that is all one person can take care of, if he does it himself. Bigger than that, you have to have partners or foremen. I am getting close to leasing 640 acres. Once I attain that goal, I would like to stay there and create a block of capital; then I intend to purchase. I chose this over buying 50 acres at the beginning and trying to make payments on it and still make a living. I didn't think it could be done. Instead, I've acquired the equipment necessary to run the operation that I have. I don't lease any machinery and the only contracting I do is for picking tomatoes.

I work like a hard-working man. I don't do all my own field work, but I do one man's worth of field work. I am the jack-of-all-trades. I am the fill-in man. I know how to do it all. Some people are excellent irrigators — they have never seen a tractor, but they can sure irrigate. And some people are very good tractor drivers, but they don't know which end of a shovel to stick in the ground. I can survey ditches, I can make ditches, I can cultivate, I can plant, I can furrow out, I can do it all. Whatever needs to be done, I do. On top of that I run my day-to-day business.

A typical day for me in the summer starts at 4:30 in the morning. I'm in the field by 5:00 to change the water. I run a couple of heads of water usually in the summertime. I come home around 7:00 for breakfast. Then it is back to the field, where I do whatever needs to be done — making ditches, setting up a cultivator, or fixing a tractor that is broke down. Around 11:30 I come in for lunch. I am usually back in the field by about 1:00. More of the same in the afternoon — planting one crop, or harvesting another, or getting set up for one thing or another. At 4:00, coffee hour, I stop. From 4:00 to 5:00, that is my family hour. I see more of my family between 4:00 and 5:00 p.m. than I do the rest of the 24 hours. It is a good time to be out of the heat. At 5:00 it is back to the field to change water and do evening work. I usually get home around 8:00 at night.

Besides myself, I have one full-time employee, my right hand man. Depending on what's needed, I hire additional help, usually two other full-time people during the summer. Sometimes I'll have as high as five or six people at one time. My foreman has worked for me the last five years. He lives in Texas. About the end of February I give him a phone call and he packs up his wife and kids and comes out here. I provide on-the-ranch housing for him. He works from March till November, then he goes back to Texas to spend the wintertime. Ninety percent of our help in this area comes from either Texas or Arizona, or from down south in the Imperial Valley. They are all Mexican, mostly with green cards. To be a farmer in this area, you have to be bilingual. I speak Spanish, because usually none of my help speak English.

The American people, in my estimation, are absolutely spoiled rotten. Very, very seldom do you find a white person who wants to do farm work. Regardless of what he's paid, he doesn't want to get his hands dirty, he doesn't want to work out in the hot sun. The Mexicans are not afraid of that, they have done it all their lives. To them, farm work is an honorable way of making a living. They are more than just laborers, they are specialists. Whereas the whites, especially the young ones coming out of the cities nowadays, think it is beneath their dignity to go out and work in the fields. Some kids still do farm work, but they are farmer's kids.

Talk about low-paying farm jobs — it's not that low-paying, at least not around here. My number-one man last year made $9,000 in the eight months from March to the first of November. I also furnished him a two-bedroom house, paid all his utilities, gave him a pickup and furnished his gas. Plus he has unlimited use of anything I have, from a shovel up to the biggest tractor.

But the white person wants to work only eight hours a day, five days a week, and wants to make $400. The Mexican-American wants to make his $350 or $400 a week too, but he's not afraid to work 100 hours to do it. So the job gets done. This overtime is a bunch of baloney, because our forefathers worked from daylight to dark. The only time they got shorter working hours was in the winter when the sun didn't shine. This country was made great, in my estimation, on that type of thinking — you do a day's work for a day's pay. Now we have the labor unions and overtime and double-time on holidays and a 40-hour work week — agriculture cannot abide by that. You can't tell plants to stop growing after 5 o'clock in the afternoon. We have to work all the daylight hours. In the fall we might work all night too. Beans, for example, must be cut at night, when there is moisture or dew on the vine, in order to keep from shattering the pods.

I try to stay out of labor-intensive crops, with the exception of the green shipping tomatoes. However, that harvest labor is contracted — I just call someone to say that the field is ready, and they bring the crew in. I scout for people to do my thinning and hoeing. I can get a contractor to come in and put

100 people in my field in three days. But generally my foreman takes care of it for me. His wife knows everybody else's wife in the countryside. They get together and say, "Hey, we need 15 people." By golly, they show up. It works quite well if you just need a few.

The trend is to get bigger. The small guy doesn't have a prayer. If you net $100 an acre, on 100 acres that is $10,000. Well, when I started farming ten years ago, I had it in my head that if I could ever make $10,000 a year I would have it made. Today I would hate to live on that, because of inflation. Yet I'm still netting only $100 an acre, maybe less. Because of inflation, I need to farm more acres to maintain my standard of living. As my margin gets lower, I have to farm more acres to maintain the same dollar value.

Acreage is available; I could get more acreage tomorrow. This farming is somewhat of a cut-throat business. I farm 450 acres, but there are only 350 that I can count on. My lease is yearly. Someone might offer to pay more rent than I do. Farmers can be pretty competitive for land. People have come to my father and tried to get ground away from me. By the same token, I really don't know anybody around here that I wouldn't go after a piece of ground he had, if I wanted it bad enough. Usually leases are two or three years. On my ground I am paying $200 an acre rent. If someone offered the landlord $210, he could take it away from me for the following year. If the landlord is honorable, $10 will not be enough for him to do that. If he is just a little bit honorable, he'll come to me and say, "Hey, I got a better offer, do you want to match it?"

I owe everything I am, and everything I have, to my father. Everything I know, the basics, I learned from him. My father was a very good teacher. He never sat me down and said, "Son, this is the way you do this." But I spent my time with him and learned by watching him. He showed me how to drive a tractor when I was young. It was 20-year on-the-job training! My father is highly respected in the area, even though he doesn't farm anymore. He always had a very good reputation for being honest and hard-working. People have a tendency to judge someone by the reputation of his parents, at least until he creates a different one. So I'm lucky.

My father's financial assistance certainly was what got me started, too. When I got out of the Marines, I had a wife and $500. How do you start farming with that? I worked as a hired man for him for a year. He paid me, and that was where I got my money to eat. The next year when I started farming myself, I didn't always have money to eat or to pay for all the things that I used. My dad made sure that I could always borrow whatever I needed, and he was also around to watch and advise. Maybe a person could start farming without family background if somewhere along the line he acquired the knowledge, but how he would acquire it, I don't know. Maybe he could learn it from books, schools, but personally I think not, because there are so many things that nobody can teach, but come only through experience. Insurance companies, production credit, banks, and savings and loan companies will loan money, but very few of them loan money to a young person just trying to get started without experience.

I see this situation as somewhat dangerous for several reasons. Some people with the potential to become good farmers never get the opportunity. As more people are retiring or getting out of farming — driven out, going broke, quitting, whatever — fewer people are coming in. Individual holdings wind up getting greater and greater. Well, anytime you place control of food production in a small number of people's hands, that is dangerous. The huge companies like Tenneco down south scare me because they are so much bigger than I am — if they wanted this place, they could get it. In this particular area there is mostly still farming from father to son and on down the line. But down south there are big conglomerates. And super-big companies are not as efficient as the smaller unit.

I hate to say it, because I want to expand, so that I can make more money — but I am not as efficient at 450 acres as when I was back at 50 acres, and I know this is true. When I had 50 acres I did absolutely everything myself, and everything was done perfectly. Today I delegate jobs to be done by different people. I check up on them, but sometimes they don't do things just right. The more you have other people do things, the less efficient your operation becomes. You lose a little bit of control. Now that I have 450 acres, I have only so much water and time and equipment. I have to choose which particular facet of the ranch that I want to take care of at any given time. If two parts both need attention at the same time, I can still only do one thing. Then I try to figure which piece will be hurt the least, and let that go, and come back and catch it in three days. But when you have only a very small acreage you can give it all number-one, 100 percent care. Your ability to care for every acre that you are farming decreases with the number of acres.

On the other hand, there is a distinction between what would be economies of scale and efficiency of management. A piece of equipment will do X number of acres, so you use your equipment most efficiently if you have the ideal number of acres it is able to service. It takes no more equipment nowadays to farm 300 acres than to farm 100. You still need the big cat, the big tractor, harvester, the cultivator, the siphon pipe. Maybe you need a few more siphon pipes and a few more tarps, but the basic equipment is there.

I like the farmer's life. I get up when I want to, I go to bed when I want to, I eat when I want to, I sleep when I want to. Nobody tells me what to do. The only thing that limits my actions is my conscience. It takes a certain sort of person to discipline himself so that he can do what needs to be done. Farmers have always, at least the successful ones, been very good at this. I like the freedom. Nobody can tell me that I can't work unless I cut my hair, or that I can't wear blue jeans with patches on them.

But what I really like most about farming is the opportunity that it gives me to teach my children a little something about life and growing up. I was born and raised on a ranch, and I would have it no other way. I want that for my kids. They can walk just about one mile in any direction and there is nothing here that is going to hurt them. How many city people can say that? My son can take his BB gun and go in any direction as far as he wants.

Last year was a very bad year for me. Somehow or other I managed to lose $46,000. The year before that was a very good year. Between what I made the year before and what I lost last year I am not very far in the hole. But if I subtract my income tax, that puts me another $20,000 in the hole. The tax system is not very just when it comes to farmers, or to anybody who has giant fluctuations in income. Income averaging helps, but the problem is that when you have a good year, you pay taxes on it, you don't get to keep it.

Farming is the only business that I can name that buys retail and sells wholesale. If we do make some money, everything we buy is with after-tax dollars. But in spite of all the problems in farming, if I had it to do over, I would do it all again.

## SHIRLEY ANDERS WEBSTER, AGE 29

I was born in 1949 in Vernal, Utah. I grew up in Bonanza, a mining camp, where they mine gilsonite, which is an ore like coal. They make paper out of it, and asphalt for roads and tiles. There were only about 50 houses in Bonanza. My father was an electrician for the mine. I went to grade school in Bonanza, but I went 50 miles on a bus every day to get to high school in Vernal until I graduated.

I wanted to see the world, so I figured the only way was to get out of Utah. I worked for a while at Flaming Gorge Lodge, a fishing resort. Then I joined the Marine Corps and came to California. I was a cook at Camp Pendleton for about a year when I met Bill. We got married, and I got out of the Marine Corps at the same time he did. We went to Utah and visited my folks, and then we came back to Grayson. We have been here ever since.

I was used to farming, actually. My dad had a dairy before he was electrician at the mine. I had been there for a while so I knew what farming was going to be. Here in California things are different, though, than in Utah.

We moved around in the beginning from place to place — from one little place to a bigger place, and then to a bigger one, and then to this one. Where we live now is not connected with the farm at all — we rent the house from somebody else. There are three houses on the home ranch: Bill's mother and father live in one, Bill's aunt lives in one, and our hired man is in the other one. This house we're in is on a separate ranch, about three miles away.

We have two children. They go to Gallup Christian School. We don't send them to the public school in town because all the migrants are there, and they don't speak English in the classroom. There were two white kids in the classroom, but not anymore — they couldn't stand it, because they couldn't speak Spanish. So most of the farmers' families around here are now sending their children to private schools. It's been about the last five years that the district has been so heavily Spanish. They teach in Spanish now, because most of the kids don't speak English. We went through the congressmen and everybody, but they just said we couldn't go to a different public school. Others have tried, and they couldn't get their kids out into another school. It is a very uncomfortable situation. It runs into money to send your children to a private school. They go clear to Gallup, about 20 miles from our house. I carpool with one other mother.

We have a family ritual. We meet at Bill's mother's house at 4:00 p.m. every day and have coffee. His mother has a swimming pool, so we have an afternoon swim with the kids. Otherwise Bill doesn't see much of them in the summer. By the time he leaves in the morning, the kids are still in bed. And then they are in bed when he comes home at night. Sometimes he drives the tractor all night. He never gets to see them then.

Most of the time I stay at home. At harvest time I take meals out to the field to feed him and the crew. Our little boy works even though he's young. If he wants to earn money, he hoes weeds in the ditches, and he starts a pipe once in a while. He is getting pretty good at irrigating even though he is only nine. Sometimes he spends the whole day with his father.

Ten years from now I hope we are more solvent than we are right now. I hope there will be a point where we can take the kids out of private school and put them back in the public school. The public high school in Sunset is probably 70 percent Spanish, 30 percent white. But by the time they get to high school, the students can speak English.

Bill wants so much out of life, and he gets a little aggravated sometimes because he is not getting there. I keep telling him that everybody else is in the same boat — just hold on, and we will make it eventually. We have had house plans drawn up for a couple of years. The first year we had to get a new tractor, so we didn't build. The next year was another new tractor. So far each year it is something else.

I never did live in town, so I don't even know what it is like. But I am glad there is not all the noise out here, like cars and horns and all the traffic. It is nice that my kids can romp around outside. The only thing I object to in farming is the hours in the summer. I don't mind anything else.

Farm family and workers stand on their mule-drawn harvester, with acres of grain as a backdrop. Reprinted with permission of the Department of Special Collections, University of California, Davis, Shields Library.

# THE RURAL ARISTOCRACY
## The Mathews Family

Over the lip of the hill and down across the small bridge spanning the Yolanda River lies the Mathews ranch, huge new dairy barn gleaming among the pastures. A cluster of trees marks the main house and outbuildings, and the hills rise again on the other side. Beautiful location, good people, a thriving business, an ordered and productive life.

The Mathews house is comfortable but surprisingly small. At 70, Ted is vigorous and articulate, a sensible, humorous individual who feels justly proud of his life's accomplishments. His speech flows smoothly, his thoughts arrange themselves logically even in spontaneous talk. He has acquired a certain cosmopolitan polish by serving as spokesman for dairymen and cattle breeders, and his intelligence and commitment are evident in every sentence.

His attractive wife Julia looks ten years younger than she is. An aura of quiet sophistication pervades her presence too. Cows may be the business of this family, but the Mathews are citizens of the world, and they know it.

THEODORE MATHEWS, AGE 70

My mother's father rode a horse out west from Iowa in 1861 in a wagon train with his widowed mother and an older brother. His father had died when he was 12 years old, and the family, with other relatives, decided to come on to California. They came to this area and our family has been here ever since. The girl that my grandfather eventually married was born in Tilson,

California, in 1865. Her father was a farmer and teamster hauling supplies to the southern mines out of Stockton. My grandfather owned his first land in this area about 1902. Previous to that he had operated livery stables in Modesto and San Luis Obispo, and worked as foreman on some of the large grain ranches. He was an excellent blacksmith and a horseman, and so he was never without a job.

In 1908 I was born in Jenning's Ford, eight miles from the ranch. The only doctor in the country was there. My grandparents were living in Jenning's Ford but farming 5,000 acres of wheat and barley. My mother went to their home for the blessed event.

When I was old enough to start school, the nearest schoolhouse was two miles south. The first year I walked to school with a couple of neighbors, except in real stormy weather when we were given the luxury of a ride in a buggy. The next year my grandfather bought me a saddle horse, and I rode to school. We broke the saddle horse to a cart when my sisters were old enough to go to school. That horse had a long life — she stayed around until she was about 30 years old.

I remember my grandfather. He taught me a lot of things. I was an only grandson, and I spent much time with him. He passed away about 1920.

The nearest high school in those days was 22 miles away. That was an impossible commute then, so I lived in Modesto with my grandmother and attended high school there my first year. School buses were started that year, so in my second year I was able to move home and commute to Riverdale High School, where I graduated. Then I went on to what was then the College of the Pacific, in Stockton. We were farming with horses and mules, dry farming, no irrigation up to that time (or very little), and I could harness and handle a team of mules as well as any of our hired help. Therefore I didn't choose to go to an agricultural college. But in this modern day, I think if a young man intends to farm, he'd better get some training in technology. My dad was often taking short courses at the University of California at Davis, and I attended a number of those sessions with him.

I had four sisters. I was the only son in the family, so there was always plenty of work. My parents were farming with horses and mules up until well into the 1920s when the first reliable tractors began to appear. My grandfather had purchased his first piece of property, maybe 500 or 600 acres, in 1902, and subsequently acquired additional grain land. The acreages may seem large, but with those kind of crops you usually try to own or rent as much land as you can handle with your teams and equipment.

Most of the time we had as many as five ten-mule teams, or 50 mules. Farming with mules was an entirely different way of life than today, because a good part of our year was spent in providing enough feed for the mules — who had to eat every day, whether they worked or not. We put up our own hay, and usually the other thing in their diet was steam-rolled barley. When mules are working, hay does not give them enough energy for a good day's work. So they were always rewarded with a good feed of barley at least once a day. Much of what we were raising was to feed the animals that we were working with — not a major portion, but it was substantial. If the whole country was operating on horse and mule power today to feed the population we have, a major output would be required just to keep the animals alive and in working condition.

We usually had one to four or five hired men, called mule-skinners in those days, meaning that they handled the mule teams. They lived in a bunkhouse and boarded right at the family table. My mother and grandmother always cooked for them. The men would harness the mules before breakfast — before daylight, most of the year. Then, after breakfast, they would immediately lead or ride the lead mule out to the field in which they were working. At lunch they sometimes would leave the mules in the field, or carry their lunch with them because of the distance. Usually we'd work until sundown and then return to unharness the mules and feed them. It was a long day by today's standards — about 12 or 14 hours.

In those days youngsters were expected to work, and we wanted to. By the time you were 12 years old, in most types of work, you could handle a man's job. The only thing that you probably couldn't do was to handle grain sacks, which might be a little more than you could lift. Wheat weighed about 135 or 140 pounds a sack, and barley about 110. Two 12-year-olds could handle them, though, and I did that on many occasions.

Our milk was all provided from one or more family cows. When I was real young, I remember well that my mother milked, because my father would be in the field until late at night. Subsequently, all through grammar and high school, I myself was milking six or eight cows by hand, night and morning, to provide milk for the family. We had a separator and we sold a little cream. We weren't considered to be in the dairy business, but it was a sideline to get a little cash in hand. Money was pretty scarce in those days. With grain, you had only one time a year when you had anything to market, usually in August. All of our bills at the grocery store in Jenning's Ford were allowed to run for one year. The grocer had to be an optimist, but he knew after harvest there would be money enough to pay. The milk money was something from day to day, or week to week. The creameries were just the beginning — motor trucks were just coming in to pick up these cans of milk. Some of those trucks were pretty hazardous, but they did get the job done. Once a month the creameries would pay for that cream which we separated. They'd pick it up daily but pay for it monthly.

My father was a great believer in diversification. He was constantly searching for something other than just dryland grain. We accumulated some livestock and eventually had a herd of registered beef cattle, shorthorns. Likewise, he put in some walnut orchards in the river bottom starting about 1917. But walnuts took years to bear in commercial quantity, and there wasn't much return on them until the end of the 1920s.

In the 1880s grain was the main crop in this state, and a great amount of wheat was produced in this area. But farmers were not practicing conservation to the extent that they should have, and the land began to suffer. Commercial fertilizer was unknown. Toward the end of the 1890s and even more toward the end of the century, they found out that they could only crop every other year. They would summer fallow, or till the land and let it lie idle for one year to conserve moisture and fertility, and then crop it the next year. Barley instead of wheat became the leading crop for many years. That was on an alternate-year basis too. Occasionally, a farmer would gamble and seed a field that had not been fallowed and rested, and if he was lucky and hit a real wet year, he could get a substantial crop. The rainfall seems to have so much in it that we don't give credit to. If we had a wet year, we could get a crop every year. Moisture, as well as fertility, was one of the limiting factors in cropping every year.

Along in the 1890s farmers began to be far-sighted enough to look for sources of irrigation water in the level areas. They found adequate water, if they had any way to conserve it, in the Yolanda River that was running off to the sea. Our local irrigation districts are prime examples of what independent people could do — they were constructed and developed without any governmental assistance whatsoever. The landowners bonded their land and amortized those bonds with the revenue generated from the electricity developed from the normal operation of an irrigation district. Our area was just out of the district because the terrain was not level, but in the 1920s my father realized that we had to do something besides dry farming. Being riparian to the Yolanda River, we developed our own irrigation system, putting pumps into the river to irrigate that land which was level enough, or could be leveled, to irrigate. This was a long time before sprinkling became popular — we used to talk about it before we'd ever seen a sprinkling system, how great it would be if that could ever be developed. My dad said many times that the hills around here would eventually be farmed. Although he didn't see much of it, it was beginning to come in. Now terrain isn't so much an obstacle whether you're going to irrigate land or not — it's soil fertility that counts.

All the hills surrounding where we are now were grain farmed. Some still are. Even the river bottom was grain farmed and yielded tremendous crops with the exception that it had a tendency to subirrigate itself. Often in the late summer the weeds would almost take over. If you didn't harvest early you had lots of problems, but the river bottom was known for big yields.

We never had a serious problem getting hired help. Many of our hired hands were sons of adjacent farmers, where they had a family with several boys. Usually one or more was able to work out for the neighbors. Likewise, there were single, itinerant laborers who more or less made the rounds. They knew when haying season was going on, and they'd show up out of nowhere. They may have been a remnant from the mining days. Some of them perhaps had come out here to mine and didn't find any gold, so they had to work. A number of those we came to know almost as part of the family. They'd stop every year for as many as 20 years in a row. They came from a variety of backgrounds. Many of them had had a tragedy in their lives in the East, maybe lost a wife at a young age. Some of them were really characters. Some would have a wagon or buggy and a horse or a burro. Many of them just came on foot and carried their bedrolls on their backs. They'd stay a month or two. Or they would come for haying, which would be in April and May, and would stay through the harvest, which would be about the first of August. The ones who did the plowing and seeding were of a more permanent type. Some lived the year round in the bunkhouse if they were single. When there was no electricity, we cooked with wood, so somebody had to cut wood, and these fellows would do that kind of work when they weren't driving a team or plowing the field. Every year after the grain harvest, there would be a period of approximately a month that we'd haul straw and store it in the straw shed. That would be the feed for the mules when they weren't working. Straw was expendable and economical, so it was their winter diet.

When we got into tree crops and more labor-intensive things, then we began to improvise. The first tree shaker that we used to get the walnuts off the trees was a windmill tower that we made right here in our own crude shop, without even an electric welder. We mounted it on one of our first small tractors. Up on the windmill tower, at three different levels, were mounted 50-gallon drums with the head cut out of it and out on an arm, so that a man could stand in there with a reasonable chance of not falling out. We would drive that tractor slowly around the trees and the man had a pole with a hook on it and a rubber mallet, and he would beat the limbs and try to get those nuts off on the ground. It was crude and inefficient compared to modern tree shakers but did get the job done. Of course, when the nuts hit the ground that's where they stayed until you picked them up by hand. It took a substantial crew to harvest a walnut crop.

In the beginning my mother, my grandmother, my sisters, and all the women of the neighborhood whose husbands didn't have walnuts would join in to pick the nuts up on a piece-rate basis. If you were a real good walnut picker you could make a little over $1 a day. I picked walnuts a number of times as a youngster for neighbors as well as home, for 5 cents a lug box. That would be about 30 pounds. I thought I was pretty good if I could pick 20 lugs a day. Eventually I saw experts who picked 100 a day. I wasn't very old when I was picking 20.

We used many Mexican nationals in the late 1940s and early 1950s. Prior to that, there were enough local people looking for work. But they disappeared, or found other employment, or didn't want to do that kind of work. In 1942, at the start of the bracero program, we used as many as 30 or 40 workers in the walnuts because we were still hand picking. We set up a facility where we fed them and arranged a bunkhouse-type dormitory where they slept. The braceros were excellent help to get the crop harvested. The walnut harvest would last about 45 days. Most of the braceros were paid by the hour, although we usually offered an alternative of piece rate. Some could double their hourly wage by working piece rate, but if they wanted they could work the hourly wage. A few were such terrific workers and so eager that we'd offer them the option and they would usually take the piece rate.

The bracero program was phased out and then we didn't have anyone. No one wanted to do that kind of work. So we began to think about mechanical picking. Some of the orchardists in the area were working on this. Many farmers were mechanics — they had to be, as well as veterinarians and everything else. We purchased our first pick-up machine in the 1950s. It was rather crude, but with a crew of six or eight we could do what the 30 had done. We began to have a hard time even getting the six or eight. Our orchard had increased in acreage and our facilities were such that it was difficult to handle enough braceros even if they were available, so we quit using the bracero program almost before it was terminated, really.

I personally have driven a walnut picker nearly every year of my life since then. When I had to acquire a new one in 1977, it was so sophisticated that I didn't have anybody I wanted to trust with it. It was a $30,000 machine. I wanted to be sure they wouldn't climb a tree with it! We harvest now with about a total crew of 10 as compared to 45 in the peak, years ago. We operate two shakers and a sweeper, a picker, a dehydrator, and a huller. No one in the whole crew works as hard as we did in the old days when we were hand shaking and hand picking. There's no comparison.

We're still harvesting those trees that my father planted before 1920. Some of them were 60 years old last year and had more production in their 59th year than they had in their entire history. We don't know for sure the life expectancy of walnuts, but if they haven't received an injury or developed a disease or crown gall or something like that, they are very long-lived trees. My father was among the first to plant walnuts in this area.

An odd disaster got my father in the walnut business. The Yolanda Reservoir, which adjoins us, was filled for the first time in 1914. Just as they got it to absolute capacity, the head gate burst out and 40,000 acre-feet of water came through our place, cutting a huge canyon 100 feet deep and 300 feet wide, and covering a good part of the river bottom land with sand, pure sand. It was a disaster; the bottom land was ruined forever. You couldn't grow anything, and you can't yet, on the surface. My father had been interested in the beginning walnut culture around us, so he drilled holes through the sand by hand and found that underneath most of the area there was still soil, although covered as deep as 10 feet with sand. With a 16-inch auger pulled up with a block and tackle by hand, he bored holes down to that black river bottom soil underneath and shoveled those holes full of dirt hauled in by a sled and three horses. And he planted young walnut trees. Having a tap root, the walnuts flourished. This is perhaps one of the most remarkable orchards in the state. We planted ten trees to an acre and they are touching solid in all directions. Many of the trunks are in excess of four feet in diameter. They're huge, huge trees, because the sand, as far as walnut culture is concerned, seemed to be just right for holding moisture. The trees were not irrigated until they were more than 50 years old. Then they began to suffer because they were so big, and we have sprinkled in the last 15 years. Those trees were a solution to what otherwise would have been a total catastrophe, but they were a long wait and a tremendous amount of expense.

I wonder, when this orchard goes, whether or not, at modern costs, we will ever be able to replant it. It would have to be fumigated, because once you have an orchard go out, whatever took its toll will probably attack the new young trees and kill them instantly. New trees are very difficult to establish on an old planting. We're losing some trees every year, but most of them from man-made injury, where we sometimes hit the crown of the tree with a disc, and then a fungus or something gets started, and away it goes. Trees that have never had an injury look as good now as they did 20 years ago, still growing vigorously. Probably the majority of the trees will go over the next ten years, but that's still pretty old.

Not having rangeland, my father disposed of his beef cattle in the 1940s. Previous to that he had purchased some dairy cattle. The dairy herd, in addition to walnuts, became one of our principal activities. Anybody who knows anything about milking cows knows that it's an activity 365 days a year. My father soon decided that he wanted better than just average cows. So in 1937 he purchased the first registered Holstein cattle. The herd was part Grade A or commercial, and part registered, from then on until about 1962, when we were able to sell the last of the Grade A. Since then we have had a 100 percent registered herd. In the late 1940s we were hoping to milk as many as 120 cows because we thought if we could, we could justify a little more manpower. That number has climbed, and now we're milking 500, besides raising enough replacements for the herd. We've got about 500 younger females, plus registered bulls that we're saving to use and sell as herd sires, and a good many steers.

The principle ingredient in our dairy ration is corn sileage, which we grow ourselves. It's an innovation that started about 1950. The whole plant — cob, stalk, grain, and all — is chopped and put into a bunker silo, a concrete slab with sides so that we can get in with a heavy tractor and pack this finely chopped corn and try to exclude all the air. In the last several years we have prepared about 7,000 tons of that material, which about gets us around the clock, 365 days, as a basic ingredient in our feed. It's a very rich and highly nutritious feed. There's plenty of energy in corn. Its fault is that it lacks protein, so we feed the animals about 15 pounds of alfalfa per cow per day, to give them the protein they need, with about three times as much corn sileage by weight. We grow enough corn for this whole operation. The alfalfa we have to purchase. With the pests we have to fight today, alfalfa has become a specialized crop. The danger of contaminating some of your other feed, or having to spray too near cutting time, makes alfalfa a hazardous business unless you're set up to handle it that way.

With dairy cattle you need a lot of help. That's one of its undesirable features. Somebody has got to be on the job 24 hours a day, seven days a week. We have cut down our labor needs through various automated systems, with three major remodels in 30 years. The first was building a new milking facility. When we were milking 120 cows per day, two men would work in that facility at once, and it would take them about four hours to milk the 120 cows, and then they would repeat that 12 hours later. We subsequently remodeled that facility and had two men again milking at once, but in two milking facilities under the same roof. We worked two shifts in one of those facilities. That remodeling lasted about 12 years. We were eventually milking as many as 400 cows. Each man would work about eight to nine hours. Three men would put in a shift in that barn. Due to the health and sanitary requirements required of California dairymen, that barn outlived its usefulness and became more difficult to maintain, so we have just recently completed a new ultramodern facility in which one man milks 64 cows an hour. It's highly automated. All the equipment in the barn is either air- or hydraulic- instead of electrically operated. There's no electricity except for lights. It's an ultramodern facility designed to eliminate the need for drugs. The barn was designed to eliminate the mastitis problem. We're of the opinion that some of the valuable drugs that we've had in treating mastitis will be removed from the market in the future.

My son Mike designed the facility with the builder. We've had close to 1,000 people here to see it in the last 60 days. We're getting interest from all over the country. We've had people from Canada, and from the state of Washington, and a group is coming from New York and Maryland and Michigan next week, totaling several hundred. The poor dairymen have had such a time over the years in a continuing fight with the mastitis problem that we hope they get a tip or two that might help.

We have worked very closely with the university for years and years — ever since we had a dairy, and even before that. We believe that the research that the university has done over the years has been a great asset to the dairyman, but more particularly to the consumer. In California we have substantially the cheapest bottle of milk in the nation. This has been possible only because the California dairyman's efficiency permits him to take a lower price for milk than other dairymen must have to stay in business. Sometimes we think it's too much of a free ride, the fact that our efficiency really accrues to the consumer rather than to the producer. On the other hand, we firmly believe that the consumer is entitled to high quality milk at a reasonable price.

California has the highest production per cow of anywhere in the nation. Only one place in the world beats us, and that's Israel. They developed a breed of cows for modern times. They had no hangups about Old Bessy, or who this line of cows was they were working with. If they didn't produce, they made hamburger out of them. The Israelis have developed a terrific producing breed of cows, primarily Holstein-Friesians. Production is the name of the game, if you're going to stay in business and make any profit at all.

In 1951 there were 20,000 dairies in California. At that time the rule of thumb was that a dairyman with 40 cows and 40 acres could make a living and send his kids through school, doing most of the work himself. Well, believe me, he had a full-time job. That is not a possibility at all anymore. When we started enlarging the herd, each time it was to justify better working conditions, or more men to do the same job, or shorter hours for our help and ourselves. When you get to using hired labor for the milking chore, which most California dairies do now, you must move the herd size by the number of cows one man and your automation can handle. If one man and your equipment can handle 150 cows, and you want to increase, you don't increase ten. He's a specialist, so you've got to increase whatever the man can handle. In our case now, with this new equipment, the milkers do not feed. The mix is automatically fed to the cows, preset as to quantity, in the milking parlor. One man in a little over six hours milks the entire herd, which is about 425 cows in the barn any given day. (With a 500-cow herd, you've got 75 in what we call their "dry" period, waiting for that next calf.)

The herd size has reached the capacity of this barn, so we can't add any more cows unless we build a new barn. Previous to moving to this one, our men were handling 120 cows each with the old automation. Many dairies in California are still between 120 and 150. The milker works twice a day, though; he's a split-shift worker. We've gone to this straight shift, so the man milks all the cows and then goes home. The employee likes it better.

The most difficult problem in the modern dairy is the securing of competent help. I put as much emphasis on the mental attitude that worker has as I do on what he can do physically. He has to want to work with livestock, dairy cows particularly, and he has to want to do a good job. There's no way you can shortcut it. Any person who doesn't have that incentive shouldn't milk cows for a living. Some of them try it, because it's one of the best-paying agricultural jobs today. But a man has to realize that he's working with live animals, and that they are smart, so he doesn't abuse them. If we ever see a man abusing an animal, we would terminate him quicker than for any other activity. We cannot tolerate it. Luckily, most men working in dairies aren't that kind, or they wouldn't be there.

A cow will hold back her milk if she's unhappy or mistreated — this thing about contented cows is no joke! It's a very serious thing. In this operation we have done everything we can conceive of to have an environment that the cows want to come into. We have put in music, partly for the men, but the cows like it too. It eliminates other minor noises and unusual sounds. The cows come in there and enjoy it thoroughly, I'm sure they do. They're less nervous. If you don't have music piped in, your milker will usually have a radio playing some loud thing that he likes. We just decided to control it, both as to volume and as to the source of the music. This Muzak has quite a variety!

We've had many Mexicans in our employ and generally speaking they're good with livestock. They're having a little trouble understanding the sophistication of the modern automation, because they have very little mechanical background. Most of the men who come here to milk haven't had much education. Mechanically, about the most sophisticated thing they've seen in Mexico was a burro.

The labor situation in agriculture is critical. Two nationalities receive special treatment in immigration, because we don't have sufficient people in this country who want to do certain kinds of work. One is the Basques, who come here to herd sheep, and the other is the young Portuguese milkers. Many Portuguese are also owners of dairies; they've grown up with cows. Many Dutch people have come in recent years, and they're hard workers. They see the opportunities that the United States still offers. Not many of our own young people want this seven-day-a-week, 365-day-a-year confinement. Even when you get up to where we are, it's very confining. We've been at it for 30 years and are now milking 500 cows, but my wife doesn't think we can get away from the ranch any easier than we did 30 years ago. Someone who has an interest has to be here pretty much all the time.

We hope that our employees will stay with us for a long time. When we've got a good man in the barn, we try to reward him. We had the misfortune of losing one of our long-time employees with a heart attack a year ago. He survived it and had open heart surgery, but he was never able to milk again. Another of our employees had been with us for about ten years and left to take a management job in another dairy. Years later he had an auto accident and was in the hospital for a long time. He lost his management job, came back here to ask for a job, and has been with us for at least ten years this time. Our foreman has been with us since 1947. We have just recently put on several new men, whom we hope have the qualities to appreciate and learn the routine and will stay with us. The living conditions and the straight shift and a number of other things, I think, are improving, and we hope will be conducive to long-term employment.

Any dairyman who goes out of business will never come back, in 995 out of 1,000 cases. Usually they go out because they're under pressure. The profit is little. The hours are long. They can't get good help even if they can afford it. They're so tired that when they finally make up their mind to go, there's no way you can get them back. People say to us when we're complaining about the price of milk, "If you don't like it, why don't you quit?" Well, it's a very difficult business to jump in and out of. The capital outlay on a dairy today is frightening. You can't buy a milk cow for less than $1,000; a good milk cow would bring more. The facilities, just the milking barn, will cost pretty close to another $1,000 a cow, yet it is worthless for anything else in the world except milking cows. Once you're in, you're a captive until you amortize the thing. It's a specialized business, and it's not for everybody.

Most dairy operations in California are still family operations. A big outfit started worrying people something terrible here about six years ago, because they were buying dairies and putting them together; they put 2,300 cows in one dairy that we know of. Everybody thought that they were going to just overrun the dairy business. Well, in five years they're down the tube in bankruptcy, because they just couldn't hire and maintain the kind of people it takes. They were too far from reality. It was an investment type thing, and it blew. Those investors wanted to make money, but the way they had to operate, any dairyman running his own business could beat them to death — and did. Dairy corporations don't last. A few have done amazingly well for a while, but generally there's got to be someone that has more than just an employee interest. A personal dedication. Working with livestock is just different than working in a warehouse or in a factory building cars.

Our son is our whole family. We're extremely fortunate that he had a knack for, and was really interested in, the livestock business. Mike is one of the better cowmen in the country — we couldn't be hanging onto this business otherwise. He's a partner with me, with a half interest in the cattle. We're in the throes of trying to develop a plan for the transfer of this operation to the next generation. Some of our advice over the years has been against incorporation, but at the moment we are very seriously considering incorporation, because acres and livestock are impossible to divide, or difficult to divide in an economical way. To hold the unit together, incorporation may be better, because you can transfer or sell interests in it. If you gave the heirs a few cows each, you'd just put them on the road to the poorhouse, for sure.

We're in a crucial time. My father passed away in 1955. My mother lived to be 91 and has only been gone two years. We have leased her land for the last 25 years. That property is right across the road and is very important in our whole dairy operation, but I have four sisters, and my mother treated all her children the same. So we have an undivided fifth interest, and eventually if the IRS ever gets through on how they're going to evaluate it, with none of the others in farming, they may expect us to purchase their interests. I hope that's what happens, to hold the unit together. I bypassed inheriting my own fifth. Under the law that goes to my legal heir, which in my case is an only child. At our age it's foolish to accumulate something because we can't take it with us. Of course, in this business, all of our profit, what we've had in the last 30 years, is out there walking around in animals.

Everybody in farming has a tremendous concern about this. The new inheritance tax laws in the last few years have addressed these issues. If we're going to keep agriculture alive and permit transfers from generation to generation (which is the only hope for the family farm), they had to approach it a little differently than a straight piece of commercial property. But when you ask IRS to appraise these properties based on their productivity, they don't know too much about the rules, apparently. They drag their feet until you just go crazy.

It's regrettable that farm numbers are declining, but economics are pretty cold and cruel. There is no way that anyone can stay in agriculture unless they expand, with their economy based on new technology. I can't think of anything more miserable than if we gave a lot of people a few cows to milk, and put them on a little family farm. They would go out of their minds very shortly.

JULIA HANLON MATHEWS, AGE 71

I was born in Amador County in 1907 and grew up in Ione, a small town. My father had farmed at one time but sold his farm and rented property and just did manual labor. After graduating from high school I went to College of the Pacific in Stockton, where I met Ted. I got a general secondary credential and was thinking about teaching school, but we were married in 1930. When he graduated, we came to the family place, a 640-acre grain ranch that his grandmother owned. We lived in an old two-story house, with flamo lights. We didn't use the upstairs part at all. Ted did trucking and hauling on the side. It was kind of lonely out there. After we had lived on the ranch for a while, and it was quite a struggle, Ted took a job with a chemical company in Modesto. We lived there then, and Mike was born there. Ted was gone usually from Sunday on, traveling extensively, and would not get back until Saturday, especially during grape season. Meanwhile I did a little substituting in high school, and then taught a couple of years in elementary school after Mike had started school.

During the war sometime, Ted's father began to want to retire, so Ted decided to come back to the ranch. During the 1930s, when farm prices were so low, Ted had some sort of a rental arrangement with his grandmother. But with the price of grain then we didn't even get back our seed a couple of years, and it was tough going. That's why he started working for the chemical company. She leased to someone else for a while and then finally sold it. So Ted was, for ten years or so, out of farming. We came back because his father knew Ted would take more of an interest in the walnut orchard and the land than anyone else, and because Ted always loved farm life. In 1946 we moved back directly into this house where we are now. His parents lived across the road, and Ted bought the land on this side.

People say it's nice and quiet in the country, but when you have hundreds of visitors, and about 15 employees, it's not so quiet. We provide housing for our employees, and we are in the center of things. Mike and his wife live up on the hill and they're a little more separated than we are from all this activity. But I see every truck and every tractor and every person who walks by. I keep track of everything, so I can alert somebody if something looks wrong. This is our maternity pasture out here, for instance, and there are times in the middle of the night when I can hear a cow bawling — you can tell by the sound that she makes, whether she's in trouble. They're just like humans, they have problems calving sometimes too. I'll call someone to help her. It's getting so complicated now that we have special people to do things: a herdsman, a calf feeder, even a nutritionist.

The office telephone is connected in the house, so when no one is in the office, it rings in the house. I take messages, and I've been sort of a central switchboard. We have registered animals and Ted and Mike have kept all of the records. They feel that they know the animals better that way, because we sell breeding stock too. I have helped with the bookkeeping, but now we have someone who does it. It's grown so complicated with all the required reports.

My husband is a trustee of the International Livestock School, which has its sessions in Texas or Arizona. He's been a director of the Bank of America Livestock Symposium, and has been in the Cattlemen's Association and the Beef Council and all of the dairy organizations. He was the national director for eight years for the National Holstein Association. So we have had many trips to different parts of the United States, many meetings and conventions and tours. We went to Europe, to Spain and Yugoslavia and Portugal, through the foreign ag service and the Holstein Association. We know people all over the United States.

This new facility we built recently has been exciting. The first evening that we milked the cows, all of us stayed up all night; we have a viewing room and you can look down on the milking operation. It was wonderful to see everything start to work together. Some of the equipment is really a first in the country. I enjoy having people come to see it. I give them something to eat sometimes, something made out of milk products. We have school classes come, from nursery school through junior college. Tomorrow afternoon we're having a class from Cal Poly. We had a group from mainland China last September. They had started at the University of Michigan and come across the country to look at different farming operations. The weather was nice, and we had lunch for them at tables all across the front lawn. There were 70 of them, including all the Chinese and the men from the Departments of Agriculture in Washington and Sacramento.

We still have a little two-room country school not far from here. It's also a community center. For years it was just a little wooden frame building, but the new one has wall-to-wall carpeting and can be opened up into one big room. About 47 children go there this year. There has been some pressure to consolidate it with a larger district. They have put it on the ballot a time or two, but people have voted no. They don't want to bus the children to the larger communities. There aren't too many country schools left. The principal teaches the four upper grades, and a woman teaches the lower grades. It has been a very happy situation. A year ago the principal decided in the spring that the children weren't learning anything inside, and they might as well have some nature study. So he and the other teacher took the entire school over to the reservoir and they camped out for a couple of nights. Some of the pupils are children of our employees — there are a number of Mexican children. They don't have a bilingual program. Some of the Mexican children have had a little difficulty when they first start, but they learn quickly. Bilingual education has been a handicap in some ways too. The rest of the children come from other ranches around.

Ted's mother and grandmother were farm people. They cooked for harvesters and were active in the home department of the Extension service. Farm women used to go to home department meetings in those days and learn to sew and to cook new recipes. That used to be women's main activity outside of church. I am less tied to the day-to-day activities on the ranch than my mother-in-law was, because I never had to do any of that harvest crew cooking or boarding the hired men. There has been increasing freedom for farm women. Our daughter-in-law is finding that even more true than I.

>Tall, pleasant-looking Mike is tired today. His responsibilities are great, and ringing phones and inquiring employees are constantly soliciting his attention. He reflects thoughtfully upon questions, but many matters are awaiting his direction, and he is glad to slip back into the all-consuming world of his dairy.

Above the nerve center of the ranch, on a bluff overlooking the river, is the tastefully appointed house belonging to Mike and Barbara, living room filled with paintings, books, and a polished baby grand piano. Barbara is statuesque, prematurely gray, a beautiful woman who is unexpectedly shy. She has led a protected life with leisure and talent enough to pursue the arts with some seriousness. Of Mike's treasures, she is clearly one of the most cherished.

MIKE MATHEWS, AGE 44

I was born in 1934. I didn't grow up on the farm; I was already 13 years old when we moved out here. I had been raised in town, and in fact we lived in Los Angeles for a while. The farm was pretty foreign to me except for visits around Christmas and Thanksgiving. I had been part of a large school in town and it was very different moving to a school with about 20 students in grades one through eight; there were just two of us in the eighth grade. I felt like it was a pretty big change in my life, but I soon was very pleased to be living here.

I attended Riverdale High School, a 20-mile ride on the bus, the same high school that my father had graduated from. My interests in high school were really quite varied, and they still are. Agriculture is not my only interest. I was never really burning up with a desire to be a farmer during my high school years. I did a little work around the farm on weekends and sometimes after school. But with that long bus ride, many times I didn't get home in time to do much work before it was dark.

The dairy on the ranch was started by my grandfather. When my dad came back they were milking about 125 cows. The dairy business was new to my father. He had been much more oriented to farming: tilling, planting, and harvesting barley and wheat. The dairy was a new ball game. I didn't get roped into doing much milking, though I can remember one time when somebody quit without notice and my father and I went out to do the milking ourselves. Neither one of us had milked very many cows at that time, but we got through it somehow or another without getting killed.

In high school I was interested in music and played an instrument and was in a dance band. That's an interest I have yet. I had a dance band in college too. I played clarinet and saxophone. My closest friends were musicians. I can remember feeling as it approached time to graduate from high school that I really had to decide what I wanted to do. I was unsure whether to study music or agriculture. I'm not at all sorry that I did turn to agriculture now, because I realize that musicians with my ability are just a dime a dozen! Even at that young age I knew that, because of the type of farm that we had, and its size and scale, I really had a fabulous opportunity to be in farming in a big way with a herd of cows.

While I was in high school one of the people that I most admired was a man named James Wandt, a veterinarian. He had the most famous herd of Holsteins in the nation. One of the keys to his success was his knowledge as a veterinarian. He was a very brilliant man. My original intent when I went to college was to get in the Veterinary School. I took the first two years in pre-vet, but I didn't distinguish myself as a great student. There were so many kids applying to vet school that I was on the bottom half scholastically and was not accepted. I changed my major to Animal Science and went straight through the four years and graduated in 1956. I met my wife in college. We got married soon after I graduated. She wanted to go to school for another year, so I did stay and work for the university for a year, but I didn't go to graduate school. As a matter of fact, I milked cows!

When we came back to the farm we were milking 250 to 300 cows and had about four people doing the milking. I started out just as a milker. I did that for several years, enough that I really know how to milk cows. Our production was quite a bit lower than it is now, about 550 pound fat average. Probably the top herd in the country at the time was about 600. We were doing an acceptable job, but we've gotten a lot better since. We're about 770 now.

The opportunity came to buy the top herd of cows owned by James Wandt, our veterinarian. Through artificial breeding we had been using the same sires for years, and our two herds were very closely related, so when he finally decided to sell his herd, it was a natural thing for us to try to buy it. His herd had been well promoted. He had taken his cattle all over the nation. Our herd, I think, was really just as good, but the difference was that his was very well known, and ours wasn't known at all. It was a great effort for my father and me to borrow the money to buy that herd, but we decided that we would, and we got publicity and mileage just out of making the purchase. We have been building on that since then. Today we have one of the highest producing herds in the nation.

When you are contemplating a big step like that, you realize that with a herd of cows, a few little things going wrong can completely destroy all that you bought. If your management ability isn't good, or if your potential customers don't have confidence in you, you cannot capitalize on that blue sky that you are buying. We did sweat about it a lot. We have never done any commercial bank financing. We have borrowed the money that we use here either from the Production Credit Association or from the Federal Land Bank. The amount of money involved, at that time, was more than our agency considered a normal loan, and they were giving us a bad time. Once we decided we just couldn't do it. But a week later we talked it over again — decided it was an opportunity that we were only going to have once, and we'd better try it. I don't regret it at all now. It was necessary.

We have only about 2,300 Grade A dairies in the whole state of California now, though there were over 10,000 in this county alone in the 1940s. As our costs like labor and feed have gone up, it has forced California dairymen to think of ways to become more efficient: to milk more cows per man, and milk them in less time. That has fostered the development of these larger herds. I was a little shocked the other day to realize that the average herd in California now is very close to 400 cows, much bigger than the national average. But the economic factors in California are different from other states. We have a favorable feed situation. California is a wonderful place to grow and buy high quality feed for dairy cows. The climate is very favorable for milk production, because most of the year cows can be outside with very little shelter. We do have some hot days. The hot days are harder on milk production than the cold days.

California imports milk, actually, in the form of cheese, but we are producing what we need in the state for the bottle. California has an unusual pricing situation whereby we have milk prices that are set by formula in a state operation. Almost all of the rest of the country has a federal order pricing system. California's prices, for various reasons — and a lot of them are political — are lower than the rest of the nation, nearly $2 a hundredweight lower than in Arizona, for example. Because of this great price difference, we are seeing small quantities of fluid milk going to Phoenix. Recently, our big milk supply out here, and the low price, have made some big-name manufacturers of cheese take a hard look at the state. Plants are being built here now to manufacture cheese. We are seeing it come very fast.

With the price of milk in this state lower than in other states, the smaller operators have been squeezed out of business. Only the really efficient dairyman can operate on those margins. I am sure that was not the intention of the marketing order in California, but that's what has happened. There is a state marketing order for milk. For about a year, a price is set by a formula that has some indexes. One of them is the cost of production — that's not a real set of figures but is based on a survey that the Department of Agriculture makes on dairy farms that are thought to be representative. Some are big, some are small, different breeds are involved and so on. They come up with some figure that is not true cost but an index. Then the Class 4 price, which is the manufacturing price, is in there;

that is related to the federal price. When the federal price goes up, our Class 4 price goes up too. Then 15 percent of the formula is the factor called Real Net Spendable Earnings, which is based on the average wages of a factory worker in Long Beach. It is some index that the state of California has been keeping for a long time for some reasons other than for this milk formula, but they use it in that formula. That takes into account the consumer's buying power — the impact of inflation, and so on.

The milk program has been a political football. It's nice for a politician to say to the voters, "Look, I am keeping the price of milk down for you." It seems to me that politics are involved in many of those decisions, because we dairymen are not very many votes. If all 2,500 dairymen voted against the governor or the party in power, it wouldn't bother them in the least.

Technological change has helped the bigger operators to be more efficient. Our computerized record keeping is one example. There is probably no business that I can think of that lends itself better to computerized record keeping than the dairy business, because we have many events that occur on predictable schedules. Heat intervals on cows, for example: We decide we want to give them so many days dry, so if you have a due date, the dry date becomes predictable. It is not like a beef operation, because we see practically every animal in our herd every day. We observe each cow, and record things that we see about her. The cow's milk is measured one day a month and tested for fat. Each time we breed an animal, it is recorded. You put it by hand on an input sheet and then have it updated once a month to a computer bank, or put it right in over the telephone line. Anything that happens we record as it occurs. If the cow calves, we put in the sex of the calf, the sire, and all of the pertinent information. If one dies, the same thing. Every day. People who work for us learn to keep close records.

The key to surviving in agriculture is to get more output for your labor. Costs have really gotten out of hand. Agriculture has the problem that we cannot compete very well in paying wages to hire the higher caliber people that we need to operate a barn like this. We need an employee who is pretty well educated.

We are quite a ways from town, and we have to provide housing. We have eight families. Most of the people who are available to do this kind of work are either immigrants or children of immigrants. Many Portuguese people grew up on a dairy, maybe in this country, or maybe their parents came here from the Azores in the 1920s and 1930s. They are comfortable around cows, but many are not well educated. They are intelligent, but don't have certain skills. Some of the automation here has overwhelmed some of them, they just cannot cope with it. When we started that new milking system last August, one milker immediately quit. It was just too much for him.

Many Mexican workers are quite willing. They have a wonderful attitude toward work and want to please you in the worst way. But some can't even drive a car. We are very limited in how we can use them — we can't give them a $50,000 tractor to run, because they will ruin it. It is a problem. We put ads in the paper when we want to hire somebody, but that is about all we can do. We don't get all that many takers. Somebody who has a year or two of college, who is interested in animals? No. Most kids today are not interested in putting in the number of hours that are required. I see this in my own kids. They are interested in the ranch, but they see how many hours I put in, and I think that my own kids — sometimes they won't admit it to me — really don't want to work that hard. Sometimes I am very happy to put in long hours, and sometimes I wish I didn't have to, but when you own the business, you do what you have to do. I am sure that I average ten hours of work a day, seven days a week. When I was younger, it was more.

When I first got out of school we were milking our cows with a bucket-type milker. You would hook the cow up to a vacuum supply, and all of her milk would go into the bucket. When she was done, you would have to take it somewhere and dump it, and then go back to the next cow. It was hard work compared to the way we do it now. One of the first changes that we made was to put in a pipeline so the milk went directly from the cow into the pipeline. We began to know more about milking equipment. I found out that most milking machines were done by the cookbook method, somebody hooked up to a vacuum pump and started using it on a cow. It was funny that they didn't improve techniques faster. We have had great advancements in the last few years in milking machines. We used to put the milker on the cow, a foot or so above the floor, and that milk would be lifted up to a relatively high pipeline. We found out that we could do a better job if the milk didn't have to be lifted, but if it could go from her udder downward. We had a much more stable vacuum when we did that. This led to the development of what we call parlor barns, where the cows are elevated and the man can milk standing up. In addition to making it easier for the man (he no longer had to go up and down), the milk could then go down to a pipeline at a lower level. That is better for the cow, and she milks faster.

We have also learned how to feed cows for higher production in the last 20 years. There are many experiment stations around the country working on milking management and feeding. We collaborate with people from the university on matters involving herd health, and have for a long time. The new equipment we have installed out here is of great interest to them, and they're doing research based on what they find here.

Mastitis is a very, very costly disease. If it is severe enough, it can be fatal. If the cow has some tissue damage, when she recovers from it her ability to secrete milk has lessened. Another type, however, which they call "subclinical mastitis," has no symptoms that you can see, and that is the one that really costs money. Less milk is secreted by the cow. You feed her for high production, but you only get part of it back.

I heard about another dairy herd where they were sterilizing the milking units in between cows. I began to read up on it, and it looked like one of the real breakthroughs in udder health in the last 20 years. A man we knew in Gallup, who eventually built this barn, started making a piece of equipment that could do this job of sterilizing the milking machine automatically between cows. When we heard of this, we were using a barn that was practically worn out anyway, and we needed to make some major changes. So we started planning the building of this new facility. We were nervous about borrowing money to buy that herd — well, we worried six times as much building this barn! It was easier to decide, though, because we could take a pencil and say if this thing works like we think it will, we are going to have this much more milk, which has this much value. We could see how we could pay for this facility, because I was pretty sure I could get the milk out of them with this barn.

We have a standby generator, which is big enough to operate the entire plant. This is pretty common in most big dairies now in California. They milk so many cows that, if they had a power failure for any length of time, they would get their schedules all fouled up. If high-producing cows go for too many hours past their normal milking time, they start to secrete less milk, and you can have problems. We don't run on that standby generator unless we have a power failure.

My father and I are in partnership. We oversee and manage the entire operation ourselves. It wasn't that way when I first got out of school. We divided the duties when I was able to take hold. I assumed management of the dairy, and he has managed the farming. I would like one of my boys to take over my position so that I can get a better idea of my dad's job all these years. The farm supplies a great deal of the feed we need for the dairy.

But my real interest is the breeding of dairy cattle. My cows are all purebred Holsteins. I like to work with registered animals. When one of the cows that you bred does something, you receive recognition for it — it's not just a cow giving milk. We are in the business of selling cows, and in the business of producing bulls. A sire analyst comes from one of these artificial insemination companies and picks out a cow that he likes, with good production. Sometimes they make an agreement with us to breed that cow to a bull of their choice. If the resulting calf is a bull, they take it when it is six months old. Many of our calves go like that, on a contract.

I think about the declining number of dairymen when I go to dairy meetings and see that there aren't very many young dairymen. I wonder sometimes if down the road there is even going to be a dairy industry. Yet the production of food is so critical to the whole world that the outlook cannot help but brighten for dairymen in the future. When it does, prices will rise, and then agriculture will attract young people again. It's a matter of supply and demand.

## BARBARA McGUIRE MATHEWS, AGE 41

My father arrived in California from Chicago in the late 1920s and met my mother at a boarding house. They both graduated from college, my father from the University of Chicago and my mother from UC Berkeley. My father took a job with California Magazine about 1930, and worked there his entire life as an editor and art director. I was born and grew up in Oakland and Menlo Park. My father's interest in art influenced me early. My mother also encouraged me in music. I started the piano when I was six, and I also had ballet lessons. I learned from my grandfather to play tennis. I was very lucky as a child to have many opportunities to learn things.

One summer in high school I worked at a dude ranch in the Sierras and met two people who were to be close friends. One was a student at UC Davis, who encouraged me to go there for college. Most of my friends went to Stanford because their fathers were professors there, but we couldn't really afford it, so I did go to Davis and met Mike the first week. I dated him almost exclusively from then on. We got married the following year.

I had two years in college. I would have liked to finish, but his father really needed Mike to come back. At that time there was no such thing as "women's lib." If I were to do it again, I would say, "Well, I will stay here and come home on the weekends until I get my degree." But in those days I didn't have the nerve.

Actually, it wasn't hard to come to a very rural area, even though I grew up in the Bay Area. I have never regretted it. I love the country. I am a person who can entertain myself. My friends thought it would be hard for me, but it wasn't. We moved back here immediately after my second year in college, and spent the summer living with my in-laws while we built this house that we are in. We have lived here all our married life.

Our first boy was born a year after we built the house. That would be 1958. Then three more boys in 1960, 1961, and 1966. For a few years I was very busy, and my husband was very busy building up the ranch. I haven't really done much on the ranch myself. There was a period after our youngest went off to kindergarten that I wanted to help. I worked about two days a week for a while in the office, but it didn't work out. Not that I wasn't happy, or Mike wasn't happy with me, but I went back to school for a while, and then started painting seriously.

Mike really needs a secretary. He is very organized and is a perfectionist. He spends a long time every day just doing the clerical work and computer entries. He works very long hours.

He talks to me about what he is trying to do on the farm. Usually I just listen because he knows so much more. He has studied and knows all the new mechanics. He was very excited about the new barn, and it has turned out well. I would listen to him describing what he wanted to do, and I knew it would be a sacrifice, but he wanted it so badly that it was fine with me.

Art is something that became a second career for me after the pressure of motherhood was off. I went back to school at a junior college. I signed up for a class with an outstanding teacher, who has been a great influence on me. Art for me now is indeed a partial career. I take it seriously, whether or not anybody else does. I don't paint every day, maybe twice a week, but I think about it all the time, and I study. I do water colors. I sell my paintings through the gallery and have also entered shows. I am fairly well known in the central California area now, and would like to continue to build up my reputation as a painter.

When I was first married I did the things that my mother-in-law did, but I found that the new friends I was meeting were mostly artists and people in Modesto. These are my close friends now. Sometimes I have guilt feelings about this. I am not as close to the farm people next door as I am to others miles away. But what is typical nowadays is not what was typical 50 years ago. Many wives of younger farmers now have come from urban areas, and farm families are more cosmopolitan than they used to be. We have friends in the turkey business, for example — she is a jeweler and she teaches advanced math to children on the side, and does other interesting things besides being a farmer's wife. We socialize with a wide spectrum of people.

I take the lead in planning recreational things. I played tennis all my life, and tried for years to get Mike to play. About five years ago he finally took a racket on a vacation to Tahoe, and he has been an avid tennis player ever since. We play about once a week at a club in town. But I have had a hard time getting him to try new things. His father worked very, very hard, and didn't take vacations. Mike could hardly remember going on a vacation with his parents, but we go every year to Tahoe for about a week. We subscribe to the San Francisco Symphony, and go to San Francisco about eight times during the season. Occasionally, if we hear about a jazz group in the Bay Area, we go there. But we really have to plan ahead to do that. Two weeks ago we went to the Modern Art Museum and saw a sculpture exhibit. We save up for big occasions in the Bay Area and don't do much locally. Once in a while we have friends for dinner, or go to someone's house.

Agriculture is definitely a man's world, and the men want it that way. I can't imagine a woman trying to get a foothold in the milk business in Sacramento — they are all men, and their attitude is that a woman should just stay out of it. Agriculture is still more traditional in social patterns than some other parts of society. The group called California Women for Agriculture is a strong and vibrant group of women, but they are basically doing things for their men, who give them the ideas.

Actually we have hired two women, both of them as calf feeders. My husband thought it would be fine to have women do that, when they came to apply. But the first one was about to get a divorce, and she would cry half the time, would have to be excused to see her lawyer, or one thing or another. She was unstable. The other girl was very good. Her brother works for us too. But it turned out that they couldn't get along, so she quit in a huff. Maybe if women go through school, get an education, and set out to have their life's work be agriculture they will make it. But they really are not prepared for ordinary farm jobs. If I had daughters I would encourage them to think about agriculture if they showed that they were really interested and willing to buckle down and learn, so they could compete with men. But they would have to be pretty determined, I think.

Our boys have worked summers and occasionally on weekends, but not doing daily chores. As they were growing up, my husband was too busy to stop and have the little ones come along, so they really didn't get into it until their early teens. Two of our sons are definitely interested in the ranch, but we have two that aren't. We are hoping that one or two of the boys will carry on, because it would be the fifth generation. We could sell the place and be millionaires and never have to work, but that is certainly not interesting to us.

When your farm has been in the family for 60 years, your roots are there. Your children learn to appreciate what their father and grandfather have done. There are advantages to being able to hike to the river and go fishing, instead of going down to the corner and maybe having a cigarette behind the fence. Generally, farm life is wholesome. Our children are more naive, in some respects, than town kids. Yet, in the long run, they are more mature. They have a strong foundation for making good decisions about life. I feel really positive about my children. Growing up here on the farm has had a lot to do with it. They see a real stable relationship between my husband and me. Maybe that would have been the same if he were in town and a lawyer or something else. But I have the feeling that marriage is even better out here — less chance of having things go wrong, or distractions. We belong to a tennis club in town, and we can see that we have been very lucky to have the life we have, when we look around at those people.

A truck farmer bunches onions in the field. Reprinted with permission from <u>A Guidebook to California Agriculture</u>, University of California Press.

# THE VEGETABLE PEDDLER'S CHILDREN
## The Yokoi Family

Vineyards and orchards greenly line the narrow roads south of the winding river. A gravel lane extends through one of the newly planted sections to an old green farmhouse. It is a plain clapboard house, but freshly painted, and its small yard is full of interesting plants.

Yoshia is a graceful woman in her late 50s. Her skin is very fair, her dark eyes gentle. She speaks of hard work and struggle, but it would be easy to imagine her in a kimono tranquilly arranging flowers. Harry is a small but strongly built man in khaki work clothes, skin weatherbeaten and darkened by years in the sun. Doubtless his view of farm workers is colored by his own experience. He knows very well the value of a day's work for a day's pay.

HARRY YOKOI, AGE 63

I was born in King City in May 1916. My father was the foreman of a large grape vineyard, and we were raised on the ranch. In 1924 we moved west of Modesto, where my dad started a truck crop vegetable farm. We moved several times, from there to Tilson, from Tilson to west of Modesto again. All through the 1920s and 1930s my father was an independent operator, renting places to truck farm here and there. At that time aliens couldn't own or rent land, so we had to use other peoples' names until I got of age. I could lease ground and I could buy, because I was born here. When I got out of high school in 1935, the family started using my name, and I started working on the ranch full time.

Times were kind of hard for the family. But in those days 50 cents would buy a lot of groceries — a loaf of bread was a nickel. Everything was real cheap. Of course, what we raised to sell was real cheap too. We raised a little bit of everything. We had a market right in Modesto — used to peddle vegetables around to each store. Safeway and another chain store called Piggly Wiggly were the biggest stores. We could sell directly to them. Now chain stores don't buy from small farmers, they buy in big quantities. But for years it was our regular business. We stocked nearly every grocery store in Modesto — about 25 or 30 of them all around town. We grew up that way. We didn't make much money, but as long as we had something to eat, we were satisfied.

My mother died in 1927 of complications with childbirth. It was a blow to my father and to the whole family. We had a rough time for a while. I was 11. My sister was in high school, so she must have been about 14. My youngest brother was just two years old. My sister stayed out of school for a year to raise the younger ones up.

I remember, before my mother passed away, in 1926 we were raising vegetable crops. I was about 10 years old. In the morning my father would harness the horse and get ready to cultivate. My mother would hold the one-row cultivator up, and I would drive the horse. My mother was real small, I don't think she was even five feet tall. In the afternoon we would pick vegetables and wash them and load them on the truck for the next day's delivery. We did everything ourselves. All my brothers would help, and my sister.

For a few years we were more or less a four-way partnership in a commune — I guess you'd call it that now, the way we farmed! With three other families we would pool our money and raise crops. When we sold them we would divide it four ways. One of the reasons we worked this way was because our parents couldn't rent land, so we had a tough time making ends meet. It was a minister of our church, who was Hawaiian born, whose name we used to rent the land. That went on for five years. Wherever we went, we rented 50 or 60 acres.

When we came to this place in 1938 we bought the land — 52 acres. It belonged to California Lands, the holding company for the Bank of America. In those days the bank got these ranches from farmers who had lost them because they were in debt so bad. The bank did anything to get rid of land, because they couldn't farm it themselves. My father got friendly with a banker and made a deal: If we could buy this ranch on a yearly payment of so much, we could get it for nothing down — but we had an agreement that we would level the ground. It had ditches going all around, but it was hard to irrigate because it wasn't leveled very well. Well, we leveled it. We hired an outfit from Tilson. We couldn't afford to level much, so we would do a piece at a time. If we made a little money, the next year we would level a little more. From 1938 until we left for the camps in 1942, everything here was vegetables. Some peach trees were on the ranch, but the variety wasn't very good. When we came back after the war we pulled the peaches out and leveled it all.

When we left during the war we didn't have much money. The ranch was rented, but our attorney would just put all the income on the payments for the land. So when we came back we didn't have anything to start work with, no machinery or anything. We gradually built up so we could buy some equipment and do hay work.

We were just as surprised as anybody with that Pearl Harbor attack. We didn't think we would have to leave our homes so quick. They gave us two weeks' notice to get out. A friend was real good to us and stored some of our furniture in his attic. The attorney took care of all the rest. We went to a relocation center in Merced at first. From there they sent us to Colorado, then let us go out to work from the camp there. We went out to hand-harvest sugar beets. We worked at it for about two months and came back with $50 each. Then the University of Colorado was offering jobs in the residence halls, so I worked there in the dining room and the kitchen. They paid $75 a month, plus board. We stayed there a couple of years. My brother and I then went to Seabrook Farms, New Jersey. They said, "If you go there, you don't have to go into the Army." Well, after they'd moved us out of our homes, we weren't about to go into the Army. Of course, when we got put in camp, they changed our draft status to a 4-C, which is "dangerous enemy alien." They never did change mine. I worked at Seabrook for about a year and a half, did a little work on the farm, and in the cold storage part. Seabrook is a big farm where they were freezing a lot of vegetables for the Army, so we had to get Army clearance to work there. We started out at 50 cents an hour and worked up to about $1 an hour. When the war was over, they let us come back.

All my brothers came back here. We were a family, everybody working together on the place. We didn't get paid or anything. Then two of my brothers left and just Jim and I stayed to farm this ranch.

The people who leased the ranch during the war put it in clover and alfalfa, and they had a small dairy. The alfalfa and clover looked good, so we got some cows and started milking them. We did that for about ten years. At that time Grade B milk was a good price so we did pretty well. But gradually it got cheaper and cheaper, and expenses started going up. We were milking about 45 cows and raising calves. We had the whole place in alfalfa and clover until 1956. Then we decided to go into some other kind of farming, so we planted new peach trees and nectarines. We still had some open ground so one year we had barley. In 1960 we planted apricots. This was all on the original 52 acres. We made the decision to go out of the dairy business in about a year's time. At that time cows were cheap, and it took a little while to get rid of them all.

The reason we decided to give up the dairy was the price of butterfat. It dropped and dropped to where it wasn't any better than going out working someplace for wages. My brother didn't care much about milking cows from the beginning. He said, "There ought to be an easier way of making a living," because it was confining — every day, no holidays. He and I did it all. We used to mow the hay, haul the hay, and work all day doing that, and then milk cows twice a day besides. It was a long day for us and not that profitable. Too many people in the business, too many improvements in the cows. Everything was going to the Grade A dairy. Butter got so high that it wouldn't sell. We didn't have any big regrets about going out of the dairy business. It was a relief. Everyplace we went, we would always have to look at our watches. "We better get home; it's time to milk."

A friend of mine who shipped fruit talked us into planting peaches. We planted Albertas, which is a fresh peach for market. He shipped them until he had a heart attack. After he retired, he was concerned about us, because he had talked us into planting these trees. He said, "Well, I will sell you my packing equipment real cheap." He told me to fix the barn up, so I did, and packed for a while, but it was too much for me. I got a contract with a buyer from Oregon, and he came down every summer and bought all my peaches. The price fluctuated. Some years were pretty good and some years they were cheap. I used to think it might not be a bad thing to go back into truck farming vegetables like we did before the war, but as you grow older and your kids don't want to farm, it's rough when you have to hire.

We started to branch out in 1967, started going out to lease, raising mainly sugar beets. We farmed out here on the Foothill Ranch for about three years. We went up to West Bend to farm there for about two years, sugar beets and tomatoes in partnership with another party. Then we got into Metro Airport. About 750 acres there at the airport belonged to the county, and we farmed it for four years — alfalfa, barley, wheat, corn, sugar beets, and melon seed crops. At the same time we got 100 acres at the Foothill Ranch and raised tomatoes for seed.

This year we raised about 300 acres of seed crops — tomatoes, melons, vegetables. You don't raise big acreages of any seed crop, because you get a lot of seed. The year we raised

100 acres of tomatoes, we got about 15 tons of seed. The seed companies don't like to have one farmer raise too much of any one thing, because if he had a crop failure, then they wouldn't have any seed. So they spread the contracts out.

Right now we are negotiating the price of seed that we are going to raise for this year. These seed companies just won't come through with what you ask for. They tell you what they are going to pay. If you don't want that, they say they'll find somebody else. You are really under the gun. I raised my labor wages this year. Everything is going up, so you have to raise wages. If you don't, they go someplace else. But for your farm products, buyers won't raise the price. It's getting to be a squeeze. The cost of new machinery is getting very prohibitive too. We bought a used tomato harvester last year. We were renting one for a while, but those rentals are so old that they are always breaking down. Well, the one we bought would break down constantly too. Then a bunch of workers would be sitting there, watching you fix it.

We have about 17 acres of peaches left on our homeplace, and about ten acres of nectarines, which we sell fresh. The peaches go to a freezer now, for pies. We had 20 acres of apricots, but that variety wasn't suited for this area, and we never did get a crop that amounted to anything until last year. We had a real good crop then, and the price was good. We figured that was the year to pull out! I had about three profitable years with apricots out of ten years.

We used to have good farm labor. There were not so many Mexican people, but plenty of families from Oklahoma were out here. They showed more pride in their work then — nowadays, all they do is come out here and put in time. They don't care about the work itself, just so they get their check at the end of the week. Once in a while you get some pretty conscientious workers, but the general quality of the workers is pretty bad. Some just sit up on the ladder, talking away, if you are not around. When you work by the hour picking fruit, if you have very many of those slack workers, the cost runs up so high that you just can't hardly make it.

With apricots we had to use some real young girls. If we didn't hire them, the family said they would go someplace else to find a job — we would have to hire the whole family. That got pretty expensive, because with some of the girls, their fathers had to move the ladders for picking. But we had to pay the kids the same as the parents. We didn't pay piece rate for market fruit, because if they work for piece rate, they pick everything, no matter how strict you are. For market, they have to handle the fruit more gently too. So we paid by the hour.

There is still no mechanization in fresh fruit harvesting. When it comes time to pick the peaches, a labor contractor comes in. Even he has a hard time getting the same people to come out and pick every day. That is the worst time of the year for us. We always have to worry if we are going to get a crew to pick the fruit. I have a small acreage, that's one reason. But another is that my contractor depends on skid-row labor — winos. Sometimes they show up and sometimes they don't — especially on Monday mornings. I have a different contractor to thin and hoe my tomatoes, and those are Mexican families who come out. In that type of work, the woman is just as good as a man, maybe better.

We have a different contractor and a different labor crew for each thing we have to do. The contractor for thinning and hoeing, his people don't like to use ladders. You can't get those people to pick fruit. They are really specialized. Those winos, I guess, like to pick fruit — most of them don't like to hoe. I guess you can call them fruit tramps. They follow the fruit around.

Seems like it would be ideal for high school kids to pick peaches during the summer, but it doesn't happen. I don't know why. Teenagers and school kids like other kinds of work, I guess. They like to drive a tractor, they don't like to pick fruit. Peaches have all that fuzz. It really irritates the picker. Actually, the local kids don't like to work on a ranch much. They'd rather find a job in town.

A farmer has to keep paying attention to what's happening all the time. Some farmers hire people and just send them out to do certain things. But it makes a difference if you are there yourself. With pruning we do piecework, and workers don't sit around then, because the more they prune, the more money they make. But you have to show your face out there once in a while, to check on the trees, because they do the least amount they can get by with. When you prune, you space the branches out so the sunlight can get in. If the farmer doesn't look around, the pruner might just cut off the top and go on.

We use a lot of fertilizer to try to get a big crop. It takes just as much time to raise a good crop as a poor crop. By spending a little more money for fertilizer, you get more tonnage in. We belong to a co-op where we buy spray material and fertilizer. We get a rebate back, so we use more fertilizer now.

I like to see things grow. But it wasn't always as much of a hassle. You knew if you raised a crop, you could make pretty good money. Now there are so many things against you. But it makes me proud if I raise a good crop. At the airport, my brother and I raised some corn that was the second highest yield in the United States. Something like that makes you feel good.

People who are not farmers should not buy and sell farm commodities. If we could sell directly to the stores, the stores could sell it cheaper. When we were kids, I remember selling carrots for a cent a bunch, about four or five carrots in a bunch. It was that cheap. Now it goes through the wholesale house and goes through so many hands that by the time the consumer gets it, it is six times what it cost to raise it. That is how everything has been. The middleman makes too much money. A friend of mine, not a farmer, bought a lot of pinto beans. He made about 2 or 3 cents a pound, and he sold them before he even paid for them. It is things like that, speculation, that raises the price of whatever you buy. The wrapper that the bread is wrapped in costs more than the wheat in the loaf of bread. When you go to Europe, the bread is not even wrapped. Everything is sold in bakeries or out in the open market. Maybe they know something we don't know. We spend too much on the outside, the packaging.

When my sons were small I worked them pretty hard. Irrigation district water runs 24 hours a day. Whenever our neighbor gets through, no matter what time it is, we have to take it. It's night work and long hours. When my sons were doing it, it turned them off, I guess, towards farming. They like teaching better. When I was a kid, my dad wanted me to farm, even though I wanted to try something else. I'd rather see my sons do what they like to do.

## YOSHIA NAKAMURA YOKOI, AGE 58

I was born in the Delta in 1921. My parents came from Japan in 1916. They stayed on Bacon Island and farmed there for a while. Later we lived in Fleming until I was about 10 years old, on a grape vineyard. My dad raised a few strawberries. We were a family of eight: two brothers, six sisters. During the Depression we moved near to Modesto, where we lived with three other Japanese families, renting and living on the same ranch down by the river for about five years. We grew vegetables, the four families together. The kids all helped, because vegetables are a lot of hand work. My husband's father was a vegetable peddler, and he would sell everything we grew. We raised almost everything under the sun: celery to lettuce, and weird vegetables like rutabagas and stuff you really didn't need. Everything you could think of. The kids used to have to help pick. We used to tie carrots, onions, almost everything, in bunches with raffia. We would throw the vegetables into the canal and the kids would jump into that water and wash them. They looked really pretty after we washed them. Every day that was our task, first thing we did after school.

The families were no relation to each other, but we all went to the same church. We separated when the lease was over. I was in high school. In 1939 I would have graduated, but I got married in 1938. So I didn't finish my senior year — I should have, but I went back and finished after the war.

My husband's family was one of the four families. Harry was five years older than I. His family didn't have a mother. When his older sister got married and left, why, we ended up getting married because he had several brothers all younger than he was. I became the mother then — I really didn't know what I was doing, I think! Two of his brothers were still in school. I immediately stepped into responsibilities. But I was used to it.

Before the war we were here for about three years. My daughter was born in 1940. I just had the one baby then, so I went out and worked on the farm too. When I first got married, it wasn't going to be that way, but I ended up working, because you can't just sit around when everybody is so busy. You have to get out and pitch in.

The war broke out in December 1941 and in May of 1942 we left for the internment camps. Before that, we kind of had the word that we were going to have to leave, but we really had only two weeks' notice. We had to get rid of our farming operation. What we did was to rent it out to a fellow. I remember it was right in the middle of strawberry season when we left. Someone else took over from the first fellow, and three years later, by the time we came back, they had turned it into a dairy.

They put us in an assembly center, they called it, in Merced. We stayed there from May to December 1942. Then they took us to Colorado, to a place called Machi. We were in that camp only for about a half a year, then they released us. The authorities would clear people to go out and work at places that would accept them. So we went to work for the University of Colorado, where they had a Japanese language school. My husband worked in the kitchen, feeding the Navy boys. He called me out a couple of months later, and I worked in the laundry, pressing Navy men's white pants with those little ironing things. It was better than being in camp. My daughter went to nursery school while I worked. I also did housework for the lady who was director of the language school. That was nice. We got to do a few things and earn a little money.

There were bad things about the camp experience, but good things too. My daughter had pneumonia twice, once in each camp. It was very unhealthy in Merced because we were in tar-papered buildings and May was so hot. I'll never forget that. The places to sleep in were really hot, and she got very sick and was in the hospital most of the time. In Colorado it snowed in the winter, and then she had double pneumonia. Camp life wasn't too good for the kids. The government did give us little supplements for the children if they were five or under — extra milk and fruit. That helped. Some of the food in the mess halls, kids wouldn't eat.

The camps were barred, so you couldn't get out. It was like a prison. There was barbed wire, and guards with guns standing at the guard posts. But there was no mistreatment. It was just that we were being held in custody. In our camp it wasn't bad. There were probably some problems I never knew about. I worked in the mess hall; we got paid $12 a month doing that.

My parents, of course, lived in this area too, so we were all in the same camps. They went to Colorado as well. My oldest sister left Colorado to go to the University of Maryland. Gradually she was able to call the parents and all of her sisters and brothers out. So they all eventually got out of camp to go to school or work. Actually, many Japanese did that. Even when they were here in California, they did what used to be called "schoolgirl work." You would stay at homes and go to junior colleges and do domestic work for room and board. All my sisters did that.

The feeling against the Japanese was much stronger on the West Coast than on the East Coast. In the East they didn't even know about it. If you got on a bus and were traveling, people would ask you all kinds of questions: "Why did they do that to you?" They didn't have strong feelings like the West Coast people. It was worse, too, in California than in Oregon or Washington. There was so much propaganda, people saying the Japanese would sabotage this and that. There were all kinds of suspicions in everybody's mind.

You really find out who your good friends are, because they would come to you — we had some who told us, "That is not how we feel — even if the public feels that way." Our neighbors, the Hollands, cooked us our last meal and stored things for us in their attic. The Nobles were very good to us too. To this day, Mrs. Noble still brings us homemade cookies every Christmas, after all these years.

Some of the Japanese did farm work in the Midwest and would live with the farm families. My husband helped harvest sugar beets in Colorado. He said it was bitter cold, and he slept in railroad cars. They had to wrap their feet in gunny sacks to keep them from freezing.

We were allowed to go home to California when the war was over, but it was still kind of dangerous to come back, because feelings were high. My brother-in-law came back first. He said the Ten Mile area was really bad. People took guns and shot at the Japanese if they saw them. I don't know how many Japanese actually lost their farms during the war. People I know still had their farms when they came back. In Modesto some of the Japanese who used to live around here didn't come back, because they didn't own anything.

One good thing came out of the evacuation: My mother learned English, because they had classes in the camp. She still can't talk very well, but she can write letters. My husband's father could already converse in English because in the produce business he had contact with people, so he picked it up. He was one of the few older ones who could talk English.

My generation never had any difficulty with English — in fact, we don't know much Japanese. That's our problem now. It's difficult to talk to the first generation, because they speak Japanese still to each other. When my mother comes, she talks to me in Japanese. I can understand her, but I answer her in English sometimes — it is all mixed. Our children, the third generation, don't know any Japanese. I have a daughter who lives in Hawaii. Her daughter goes to Japanese school, so she will probably end up knowing more of the language than her parents.

Our children all went to Modesto High School, and they all went on to college. They are all public school teachers now. My daughter teaches fourth graders in Hawaii. Jerry is an athletic director — he coaches football, golf, and wrestling, at a high school in the Kensington district. Dave teaches business courses, and coaches football, wrestling, and track. My husband and I did not encourage them to go into farming. They made up their own minds about what they wanted. My husband is in partnership with his brother, and this place isn't big enough to take care of another family.

I have always been active. I used to pack fruit in a packing shed when the kids were little. Seasonal work, about six weeks. Then I worked out when they were packing rations for the armed services during the Korean War. And I worked in a frozen food plant for a while. I did all kinds of things. When the kids were little I couldn't work full time, and I really didn't have the education. In high school I took college prep courses, but not typing. When I finally went back to school I took typing and shorthand and got a job at our local winery in 1955. I have been there ever since. I worked full time until about six years ago when I had a brain tumor operation and then I had to slow down. I work only part time now. I do media work. My boss buys television and radio spots for advertising wine and I work with him. I like to get off the farm — if you stay here, you have to work anyway and don't get paid for it! I helped send all my kids to college. I used to pick strawberries for a neighbor, to make pin money. I used to pick grapes for my neighbors. But if you get a soft job like sitting in an office all day long, it is much easier!

My parents were always education-minded. When I got married at 17 they were unhappy about it. All my brothers and sisters have college educations. I am the only one who didn't. None of my brothers and sisters have stayed in farming, either. Out of eight of us, there is not one in agriculture. One sister did study soils; she got her Master's in that.

My husband's family is all boys. The youngest is an electrical engineer. He is the only one who went to college. Of the rest, one works in a nursery, one has a gardening business, and my husband's partner is his brother, so there are two farmers. Of course, the nursery business and the gardening are working with growing things too. If you didn't have the education to do anything else, why, that is what you did. They grew up watching their parents doing the same thing. When my parents' generation came here from Japan, they really had to work hard when they first came. They wanted the kids to get off the farm, because it was such a hard life.

My husband is about ready to retire. He is 63 this year. Another couple of years and he will get social security. It probably will be easy to lease the farm out. He is converting it all into almonds now, and putting in an underground irrigation system. All you have to do is turn a knob, and the water comes on. He is getting too old to irrigate — at night you have to track through the orchard.

My son helps during the fruit season in the summer. He has a degree in accounting and he does all the payroll. He knows how to do the manual part of it too. If their dad needs an extra tractor driver, he will always call the boys, and they will work for him. Who knows, if Dave gets the ranch after we die, he might want to farm it. Meanwhile, he has a profession.

There isn't much a wife has to say about farming — especially me, because I don't know that much. My husband says, "That is because you don't have enough interest in it, that's why." All my life, it seems like every time we had extra money, we were putting it into the farm. We were always buying equipment, we never got anything for the house. Finally he decided, "Yes, we could remodel our kitchen." When we bought this house 40 years ago, it must have been 50 years old already. It was a weird house, with only one bedroom. We had to fill in the porch to make bedrooms for the kids. Last year Harry built a new kitchen with the help of a carpenter down the road. That was one of the first times that money hadn't gone back into the farm.

Last year in May I took a trip to Europe for three weeks and left him. He couldn't leave that time of the year, and he really doesn't like to travel anyway. I know he wouldn't have liked some of the things we did. He doesn't like to go to museums or look at statues and castles; that's not for him. Probably the only part he would have enjoyed was seeing the little farms. We rode on the rail most of the time, through the countryside. He would have enjoyed that part.

We have been married 40 years. It has been a pretty good 40 years. As you grow older, you appreciate it more. In my younger days I used to wish to get off the ranch, to live in town where there wasn't all the dust. Now I'm glad I don't have all that traffic and noise in town. It doesn't bother me that our sons aren't going to farm. I think, though, that by the time they are ready to retire too, they will be glad to have a farm to come back to.

This letter from 81-year-old Shizuka Nakamura was received by her daughter in January 1977. In broken English she sums up fragments of her life and expresses her intense religious devotion.

My dear Children,
Today I want to tell you what I believe is most important and thankful. First of all, it is the testimony that Our Heavenly Father led us well by his strong and kind hands. I ask you my excuses first.
I married and came to America without any consider, under the suggestion of my parents and brother. Fortunately, your father was a gentleman that loved goodness and justice and had generous nature. Therefore, he was very kind to me, though I was so selfish and weak both physically and mentally, while our children were brought up well. I am very sorry that I couldn't show my love to you enough because of my weakness and busy work, and even gave you rebukes without understanding. However, I believe that I have loved you more than my life.
But, I might have taught you more lovely.
At last, I repented of all my sins before Jesus Christ, and believed that Jesus dies on the cross to save me. Since I was saved, I talked to Him in prayer and He helped me.
Oh dear my children don't you remember that happy life in Tilson, California, for ten years. We used to sing a Japanese hymn all together in the morning home service:

> Iewa kinno naki Azora no moto,
> Mizu wa kawakanu Inochi no Mashimmazu
> Kate wa Yutakani Amakudaru mana
> Aa Bulah! Waga chiyo.

### Dwelling in Beulah Land

Far away the noise of strife upon my ear is falling,
Then I know the sins of earth on every hand—
I'm living on the mountain underneath a cloudless sky,
Drinking at the fountain that never shall run dry—

In the evening, you used to preach and sing by turn as if you attended the Sunday school with cheers. Your father merrily pretended a horse and walked around riding you on his back, didn't he nice? After all papa and mama spoke a little and prayed for you, before we went to bed. You had a habit praying by bed. Our home looked like a small heaven, for nobody spoke evil of others and dirty languages. We had a great hope that it was so happy if you had kept the pure faith and grown up. But the time passed, and our grape farm washed away by just came of the big depression wave so we had to move to Modesto in 1931.

We three Christian families as very poor we were, worked all together, only trusting in God. Some of you were senior high school students, some one were in grammar school, but you worked in the field till it was dark, as soon as you came back from school. Even little children washed vegetables with hands which were red with cold water in winter. How often mama burst into tears in the shadow, clasping my hands, when I saw you working so hard!

The pastor always had intercession for us at that time and loved to lead us. We heard that they called our farm, which included one more home and three old men, a Christian farm. There were living 35 fellow believers. Children used to sing hymns on the truck on the way to church every Sunday. Though we were so busy and poor, I lived in the confidence that the Lord was with me and kept working in prayer. I was given the power from above in His presence even while I was almost going to come down. I believe every one was the same experience in our farm. Our little church was quite like a home, and we loved one another, we were much blessed in every service. It was the time of blessing in particular when we had the evangelical meeting every year.

For us adults, it was very good but I am afraid that for you children, it was too hard, because there were no time for playing and study also. I felt very sorry to you. I thank Lord He protected us in those days. The Gospel which was preached there was true indeed and all those who believed in Jesus Christ and cross not only will be saved but cleansed.

Six of you had faith and were baptized, and Frank went to army. At the same time, the war between America and Japan began and we entered camp, so we couldn't have much chances to talk one another because of our distance. Thank you very much, Miye! For you went boldly to Maryland University at first, and called there your brother and sisters one by one. At last, you invited your parents and youngest sister to look after. To my gladness all our children helped one another with one accord.

By the grace of God you were brought up by far better than me. You all in particular were given your best half and have sweet homes with pretty lovely children born now. I am sorry before God and you that I couldn't do my best as a mother and your good example, as a Christian. However, I have looked on Jesus and went forward until now. Jesus Christ made me be reborn and gave me a new heart. He is now renewing my heart day by day. If I hadn't been saved by Jesus, I should have been in more miserable condition, the peace of our home and you had been broken bitterly. By the mercy of God we are what we are.

I have nothing to leave you, but the only precious and eternal life that Jesus Christ gave you and me. Remember the words of God that were taught you in your boy and girlhood, and have confidence in God again please.

When you presented gifts to me on mother's day, I was so thankful, thinking your kindness deeply. I want to live quietly the rest of my life in the service of God and men, and pray for you. I don't want to give you troubles more as possible as I can.

We have distance partly now, but I don't feel much lonely when we believe in Jesus and have fellowship in prayer as a true family in spirituality, because we are always one in hearts. Please, read this my letter and give my best regards to your children. May God bless each of you more abundantly and forever.

Amen.

Your mother

David Yokoi is third-generation Japanese and a member of the professional class. He has the gentle look of his mother, the strong body of his father. There is no sentimentality in him about farm life. He thinks the life of a teacher is clearly more comfortable and more rewarding. In his comfortable suburban home, he seems slightly puzzled that anyone should ask.

## DAVID YOKOI, AGE 34

I was born in 1944. I grew up with a dairy on the farm. My brother and I used to help feed the calves and milk the cows when we were about 10 or 12. After the cows were sold, we had an orchard. Then my dad and uncle started raising sugar beets and tomatoes and other row crops on leased ground. We worked in the summers on the farm, my brother and I, until we went to college. We would help fertilize or irrigate the trees. My dad would pay us.

I never did take any ag classes in high school. My parents wanted us to get an education and to try something else. They are doing better now in farming than they were then. My dad has expanded so much — before, he used to be on just 50 acres. The last couple of years, they have rented close to 1,000 acres. It is more profitable now.

I don't think it ever crossed my mind to stay on the farm. My interests didn't really lie that way. We always knew we were going to college. I majored in accounting during college, then went into business education. I worked for an accountant for a while, but I didn't like all that pressurized work, so I am a teacher now. I still work for my dad in the summer. It's a change from teaching, the physical labor. I enjoy that then.

But I don't have plans to take over dad's ranch when he retires. My family likes town life. I would rather see my dad sell his land when he retires, if he really doesn't want to farm anymore. But he seems to be in good health. He is planning his farming so that he won't have to do as much physical labor in the future. I don't think he could live in town. He likes to be out there.

# Five Years After

The interviews on which this book is based were taped in the late 1970s. Half a decade later, as the book goes to press, what changes have taken place within the families, and in the counties? Adjustments have been incremental, but continuous.

Steve Dietz now grows 150 acres of tomatoes on rented land. He is proud to be solvent — "I don't owe anybody anything" — even though 20 percent of California's farmers are said to be in financial trouble. He still does most of his own field work himself because, he says, "I can't find anyone who can do it as well as I can — but I can't expand anymore either, because I'm up to capacity!" His father Herman, at 87, has "slowed down" but still does chores. Steven's oldest son, a graduate from Stanford in industrial engineering, works in the computer industry until such time as he might be able to farm with his father.

Though Colusa County now has a tomato paste factory to augment its farming economy, other changes have been undramatic. Canal water has become available to farms on the west side, and row crops and orchards are increasing year by year. Particular expansion has taken place in almonds, where young trees produce twice as much as older ones. Like nearly all California counties, Colusa has seen a steady increase in minority population, primarily Spanish-speaking, and its unemployment and welfare rates are higher than in years past.

Helen and Pat O'Leary have moved back to the 100-year-old Elk Valley ranch, into a new home within a mile of the old homeplace where one son now lives. They have incorporated their cattle operation as "O'Leary Brothers," and Pat is ranching cooperatively with his boys.

The Breidenbachs lost grandfather Wilhelm at age 85, but Dan has married and added a new son to the family.

The Weidner family was struck with tragedy when John at age 60 was killed in a small plane crash over the Sierras, but his sons have taken over the reins in accordance with well-thought-out family plans. Buck and Laura have produced a daughter to make the fifth generation on that farm.

Tom and Genevieve Savely continue to live, retired, on their farm. Roberta is working in the San Francisco Bay area as a tour director, while Ed has become an airline mechanic elsewhere.

In Stanislaus County, land use is still a hotly contested local issue. Because of the great inflation in land prices, some farmers as well as developers are pushing for smaller lot splits.

Acquisition of additional farmland, except in small chunks, has become almost prohibitive for some farmers. While Modesto has restrained its growth through public policy, smaller communities throughout the county are experiencing the same phenomenal pressure to develop, as industries like electronics search for new sites to settle. Stanislaus, sometimes referred to in years past as "Santa Claus" county because of its liberal public assistance programs, has attracted numbers of farm workers who settle out of the migrant stream but depend on unemployment compensation in winter. Meanwhile, farm advisors say that many of the specialty commodities represented in the county have been plagued with overproduction; some farmers are taking a pounding as prices stay low while costs continue to escalate.

Homer Lind has two sons farming with him now, and he devotes part of his time to duties on the county Board of Supervisors, where he is much involved with land use issues.

While George Lowe has somewhat reduced his poultry operation, his sons assist him with their jointly owned almond orchard, now coming into production as an excellently managed operation.

Though they threaten to quit every year, the Schoppes still grow boysenberries. One son has joined Bob on the dairy and ranch, while grandfather Walter enjoys activities as a community patriarch.

Bill and Shirley Webster have been able to build the home they wanted, and there is a new baby in their family. Bill is active as a local spokesman for the Farm Bureau.

The Mathews dairy is one of the outstanding dairies in the state and nation. Ted has gradually retired, while two of Mike's sons have decided to farm with him.

Harry Yokoi is in his last year of farming with his brother, with an eye to being an orchardist — "a gentleman farmer" — when he retires from the seed business next season.

And so, in some families, the mantle has passed.

And from the planted seed, across the tilled brown fields, January rains bring forth the wheat, a shining and incredible green. And from the base of sturdy plants, shoots spring up, enriching the field. And as the months march by, the tillered wheat endures and thrives, nurtured by careful hands to harvest, bounty and beauty mixed in heavy heads of nodding grain.

January 1983

# Epilogue

Across the patient and uncomplaining land, generations of human beings come and go. The rhythms of the land are long and slow. In its native state, change comes gradually, incrementally, as seas and rivers rise and fall, rocks crumble into sand, vegetation spreads by wind and weather. Earthquake, storm, flood, and drought all have done their share to shape the landscape in California, but by degrees. The early California Indian tribes walked lightly across the land and left few tracks. Then came the Europeans and the Americans, and the country began to change, slowly at first, then more and more rapidly. The fertile Central Valley, full of tall grass, and ebbing and flowing with life as the winter rains and melting mountain snows provided means, has been harnessed, channeled, and controlled by the descendants of those early settlers. From an airplane flying high over the valley now, one sees an immensely productive and carefully contoured irrigated agriculture laid out in variously geometric and greenly textured fields, contrasting with the waterless brown and yellow rolling hills. The landscape is dramatically, radically altered for man's purposes, and it has happened in less than a hundred years, after so many centuries of silent evolution.

Human clocks, set to varied measures, are ticking now on the California land. Farming families reproduce themselves, and sons step into the footprints of their fathers. Education changes ways of viewing things. Political elections influence decisions on land use. Economics impels other kinds of decisions. Inventive, restless man tinkers with the short-term inefficiencies of nature, and reaps surprising rewards of fecundity and abundance with new technologies. Pressured by population growth and his own rising expectations, he carves out new landscapes, redirects the water, guides the genetic destiny of plant and animal species, and puts together ingenious machines to do the work of human muscle. What a clever, creative animal is man, and how justly proud he should be of his own amazing accomplishments! He has created new forms of plenty, and of beauty too, where he has turned his skilled and caring hand.

But other clocks are ticking too. Some of what men do lingers on long after the immediate effects are known, and in this interconnected world change sets off a string of aftereffects not always expected or intended. Dislocations in nature and in human society unexpectedly appear years after change begins. Some kinds of technological change turn out to affect the environment disastrously. Land that should never have been tilled begins to erode away; the poisons in pesticides seep into water supplies or threaten the reproduction of wildlife; soils mismanaged become saline and unusable after only a few decades. Social change on the heels of technology may disrupt the old protective mechanisms that kept communities stable and individuals satisfied in simpler times.

The story is always mixed, and it is not a fable with a moral. Science and technology have led in the twentieth century to an unprecedented accumulation of physical power to transform resources; as never before in history, man has left his mark on the planet. Science and technology have lessened some suffering, broadened opportunities, and extended and enriched human life. Yet the seemingly inevitable economic consequences of technological change have also had disturbing effects on old value systems, forcing a reexamination by thoughtful people of their own expectations and desires. The thrill of new options is balanced in part by regret for the passing of old ways.

Our uncertainties about change exist because science, technology, and economics are value-free; they stop short of the questions men have asked themselves from time immemorial: Why are we here? What should we live for? How shall we live? And these are still the essential questions.

# About The Author

ANN FOLEY SCHEURING is a writer and editor specializing in agricultural subjects. Since 1975 she has worked on a variety of special projects with faculty and staff of the University of California at Davis.

She has published a book-length biography through the Oral History Center at the Davis campus, in addition to articles and monographs. Her most recent books include <u>A Guidebook to California Agriculture</u> and <u>Competition for California Water</u>.

Mrs. Scheuring holds an M.A. from the University of California at Berkeley, and an M.Ed. from Davis.